Conference Proceedings of the Society for Experimental Mechanics Series

Series Editor

Kristin B. Zimmerman, Ph.D.
Society for Experimental Mechanics, Inc.,
Bethel, CT, USA

More information about this series at http://www.springer.com/series/8922

Sharlotte Kramer • Jennifer L. Jordan • Helena Jin • Jay Carroll
Alison M. Beese
Editors

Mechanics of Additive and Advanced Manufacturing, Volume 8

Proceedings of the 2018 Annual Conference on Experimental and Applied Mechanics

Springer

Editors
Sharlotte Kramer
Sandia National Laboratories
Albuquerque, NM, USA

Helena Jin
Sandia National Laboratories
Livermore, CA, USA

Alison M. Beese
Pennsylvania State University
State College, PA, USA

Jennifer L. Jordan
Los Alamos National Laboratory
Albuquerque, NM, USA

Jay Carroll
Sandia National Laboratories
Albuquerque, NM, USA

ISSN 2191-5644 ISSN 2191-5652 (electronic)
Conference Proceedings of the Society for Experimental Mechanics Series
ISBN 978-3-030-06972-8 ISBN 978-3-319-95083-9 (eBook)
https://doi.org/10.1007/978-3-319-95083-9

Softcover re-print of the Hardcover 1st edition 2019

This Springer imprint is published by the registered company Springer Nature Switzerland AG
The registered company address is: Gewerbestrasse 11, 6330 Cham, Switzerland

Preface

Mechanics of Additive and Advanced Manufacturing represents one of eight volumes of technical papers presented at the 2018 SEM Annual Conference & Exposition on Experimental and Applied Mechanics organized by the Society for Experimental Mechanics and held in Greenville, SC, June 4–7, 2018. The complete proceedings also includes volumes on *Dynamic Behavior of Materials*; *Challenges in Mechanics of Time-Dependent Materials*; *Advancement of Optical Methods & Digital Image Correlation in Experimental Mechanics*; *Mechanics of Biological Systems & Micro- and Nanomechanics*; *Mechanics of Composite, Hybrid and Multifunctional Materials*; *Fracture, Fatigue, Failure and Damage Evolution*; and *Residual Stress, Thermomechanics & Infrared Imaging, Hybrid Techniques and Inverse Problems.*

Mechanics of Additive and Advanced Manufacturing is an emerging area due to the unprecedented design and manufacturing possibilities offered by new and evolving advanced manufacturing processes and the rich mechanics issues that emerge. Technical interest within the society spans several other SEM technical divisions such as composites, hybrids and multifunctional materials, dynamic behavior of materials, fracture and fatigue, residual stress, time-dependent materials, and the research committee.

The topic of mechanics of additive and advanced manufacturing included in this volume covers design, optimization, experiments, computations, and materials for advanced manufacturing processes (3D printing, micro- and nano-manufacturing, powder bed fusion, directed energy deposition, etc.) with particular focus on mechanics aspects (e.g., mechanical properties, residual stress, deformation, failure, rate-dependent mechanical behavior, etc.).

Albuquerque, NM, USA — Sharlotte Kramer
Albuquerque, NM, USA — Jennifer L. Jordan
Livermore, CA, USA — Helena Jin
Albuquerque, NM, USA — Jay Carroll
State College, PA, USA — Alison M. Beese

Contents

1 **Structure/Property Behavior of Additively Manufactured (AM) Materials: Opportunities and Challenges** 1
George T. (Rusty) Gray III, Veronica Livescu, Cameron Knapp, and Saryu Fensin

2 **Fatigue Characterization of 3D-Printed Maraging Steel by Infrared Thermography** 5
Corentin Douellou, Xavier Balandraud, and Emmanuel Duc

3 **Quasi-Static and Dynamic Fracture Behaviors of Additively Printed ABS Coupons Studied Using DIC: Role of Build Architecture and Loading Rate** 11
John P. Isaac and Hareesh V. Tippur

4 **Compression and Shear Response of 3D Printed Foam Pads** 21
Wei-Yang Lu

5 **Mechanical Structure-Property Relationships for 2D Polymers Comprised of Nodes and Bridge Units** 25
Emil Sandoz-Rosado and Eric D. Wetzel

6 **Mechanical Behavior of Additively Manufactured Ti-6Al-4V Following a New Heat Treatment** 29
Jonathan P. Ligda, Brady G Butler, Nathaniel Saenz, and James Paramore

7 **Dynamic Thermal Softening Behavior of Additive Materials for Hybrid Manufacturing** 31
Steven Mates, Mark Stoudt, Gregor Jacob, Wilfredo Moscoso, and Vis Madhavan

8 **Correlation Between Process Parameters and Mechanical Properties in Parts Printed by the Fused Deposition Modeling Process** 35
Samuel Attoye, Ehsan Malekipour, and Hazim El-Mounayri

9 **Mechanical Characterization of Cellulose Nanofibril Materials Made by Additive Manufacturing** 43
Lisa M. Mariani, John M. Considine, and Kevin T. Turner

10 **Shock Propagation and Deformation of Additively-Manufactured Polymer Foams with Engineered Porosity** 47
Jonathan E. Spowart, David Lacina, Christopher Kit Neel, Geoffrey Frank, Andrew Abbott, and Brittany Branch

11 **Mechanical Characterization of Fused Filament Fabrication Polyvinylidene Fluoride Printed (PVDF) Composites** 59
Niknam Momenzadeh, Carson M. Stewart, and Thomas Berfield

12 **Influence of an Extreme Environment on the Tensile Mechanical Properties of a 3D Printed Thermoplastic Polymer** 67
Jose Torres, Otito Onwuzurike, Amber J. W. McClung, and Juan D. Ocampo

13 **A Framework for Estimating Mold Performance Using Experimental and Numerical Analysis of Injection Mold Tooling Prototypes** 71
Suchana Jahan, Hazim El-Mounayri, Andres Tovar, and Yung C. Shin

14 **Effect of Processing Parameters on Interlayer Fracture Toughness of Fused Filament Fabrication Thermoplastic Materials** 77
Devin J. Young, Cara Otten, and Michael W. Czabaj

15 **Forced-Response Verification of the Inherent Damping in Additive Manufactured Specimens** 81
Onome Scott-Emuakpor, Tommy George, Brian Runyon, Bryan Langley, Luke Sheridan, Casey Holycross, Ryan O'Hara, and Philip Johnson

16 **Computational and Experimental Characterization of 3D Printed Components by Fused Deposition Modeling** 87
Koohyar Pooladvand and Cosme Furlong

17 **Linking Thermal History to Mechanical Behavior in Directed Energy Deposited Materials** 97
Jian Cao

Chapter 1
Structure/Property Behavior of Additively Manufactured (AM) Materials: Opportunities and Challenges

George T. (Rusty) Gray III, Veronica Livescu, Cameron Knapp, and Saryu Fensin

Keywords Additive manufacturing · 304 L SS · Microstructure · Mechanical behavior

1.1 Scope

The certification and qualification paradigms required for additively manufactured (AM) metals and alloys must evolve given the absence of any broadly accepted "ASTM- or DIN-type" AM certification/qualification processes or fixed AM-material produced specifications. This is in part due to the breath of the evolved microstructures produced across the spectrum of AM manufacturing technologies including powder bed and directed energy systems. Accordingly, design and microstructure optimization, manufacture, and thereafter implementation and insertion of AM-produced materials to meet the wide range of engineering applications requires detailed quantification of the structure/property behavior of AM-materials, across the spectrum of metallic AM methods, in comparison/contrast to conventionally-manufactured metals and alloys [1–5]. The scope of this talk is a discussion of some present opportunities and challenges to achieving qualification and certification of AM produced metals and alloys for engineering applications.

1.2 Opportunities

AM of metallic components, as has been detailed by a number of recent reviews [1, 2, 4, 5], can offer opportunities for manufacturing components whose design can be topologically optimized into components that can neither be cast nor readily machined. In addition AM can lead to increased energy efficiency, cost, factory footprint reduction, time savings during manufacturing, and significant material savings for a number of low-volume-high value applications, plus the potential of increased yields from starting feed materials to finished components or subassemblies [1, 2, 4, 5].

1.2.1 Alloy Development

AM to date has been principally focused on utilizing either pure metals or existing engineering alloys available in powder or wire form. Ti-6Al-4 V is a good example of this but so are 2000, 5000, 6000, and 7000-series aluminum alloys. Ti-6Al-4 V is an alloy developed in the 1950's for direct cast applications and/or where an initial cast billet is followed by thermomechanical wrought processing to produce product forms and thereafter machined to produce components with optimized microstructures to meet strength/ fracture toughness/fatigue requirements, in particular for aerospace/defense applications. The compositions of many wrought Ti- and Al-based alloys were optimized to achieve the desired microstructures for either the direct cast or as a prelude to wrought processing following solidification rates typical for cast billets. However, the rapid solidification rates in

---

Los Alamos National Laboratory (LANL) is operated by Los Alamos National Security, LLC, for the National Nuclear Security Administration of the U.S. Department of Energy under contract DE-AC52-06NA25396. This research was funded by the LDRD DR-20170033 program at Los Alamos National Laboratory.

G. T. (Rusty) Gray III (✉) · V. Livescu · C. Knapp · S. Fensin
Los Alamos National Laboratory, Los Alamos, NM, USA
e-mail: rusty@lanl.gov

S. Kramer et al. (eds.), *Mechanics of Additive and Advanced Manufacturing, Volume 8*, Conference Proceedings of the Society for Experimental Mechanics Series, https://doi.org/10.1007/978-3-319-95083-9_1

laser-powder-bed AM as one example, often in the 10^6 °C per second, are either far too fast or the solidification rates in high-vacuum wire-fed electron beam DED AM processing techniques are far too slow. Alloy development for AM processing cooling rates (both very fast in laser-powder-bed AM favoring atypical microstructures or very slow in wire directed energy deposition AM machines favoring large columnar epitaxial growth yielding poor fracture and fatigue properties) and inert build AM atmospheres enhancing interstitial pick-up affecting properties, each suggest a critical need for funding in alloy design for AM. An example of this need is addressed in a recent publication where a high-strength aluminum powder was functionalized to seed grain nucleation by adding nanoparticles of nucleants, to reduce or eliminate gross epitaxial growth by controlling solidification during AM processing [6].

1.3 Challenges

AM production of metallic components must be bracketed within the scope of the relatively high expense of AM machines, high feedstock (powder or wire) costs, limited build volumes and dimensions at present, slow deposition/build rates, in addition to the evolving certification/qualification procedures/requirements for critical applications [1, 5].

1.3.1 Advances in Powder Production: Chemistry and Cost Reduction

While the cost of cast or wrought alloys of widespread structural industrial importance can vary widely, the cost of powder or wire to support AM production currently ranges from ~2× to ~an order of magnitude more than wrought or cast product at present for many metallic materials. For example, while wrought 304 L SS is currently $6–10/kg, pedigreed 304 L SS powder is ~ $85/kg and while wrought Ti-6Al-4 V is ~$40–50/kg, current ELI-grade Ti-6Al-4 V powder is $300/kg. This substantial cost in the feedstock materials needed for AM, in addition to the substantial financial investments required for AM machine purchase, let alone the operating costs, significantly restricts the market for AM production. As such AM remains centered principally on high-value – low volume component manufacturing where AM can offer unique fabrication approaches not readily available or impossible to produce via direct casting and/or reductive techniques from existing feed stocks. Accordingly, research investment in optimized /high volume and yield/reduced cost powder production technologies is a promising opportunity that is crucial if AM is to be expanded from its current restriction to low-volume high-value component manufacturing applications.

1.3.2 AM Technologies Produce a Spectrum of Microstructures

Given the differences in feedstock employed and the melting and solidification rates between laser versus electron-beam and powder-bed versus DED techniques, it is not surprising that a spectrum of microstructures can be developed by different AM platforms [1, 2, 4]. An illustration of the spectrum of microstructures and properties produced by laser or electron-beam powder-bed, directed-energy-deposition (DED) techniques such as laser-directed energy powder fed or electron-beam wire fed machine AM produced 304 L SS in comparison to wrought is presented in Fig. 1.1.

Electron-backscatter diffraction (EBSD) images of the microstructures from an ASTM-spec. wrought plate of 304 L SS and the microstructures of identically sized AM-304 L SS plates produced on four different AM machines utilizing identical 304 L SS chemistry powder are given in Fig. 1.1. The microstructures of the wrought versus the 4-AM 304 L SS plates are seen to be significantly different. While the wrought 304 L SS displays an equiaxed-polycrystalline grain structure, the microstructures of the four different AM produced 304 L SS display a spectrum of microstructures in terms of both length scale as well as the degree of morphological anisotropy. With the exception of the ARCAM build, each of the other three AM builds exhibit anisotropic elongated larger grain structures aligned with the long axis of the grains parallel to the AM build direction consistent with the solidification direction during the build. Although structure-property data probing many PSPP relationships is important in qualifying AM products, a major concern remains when comparing AM parts with those made using traditional metal manufacturing methods, in particular the failure and damage mechanisms that arise from the unique microstructures produced during AM.

Further, efforts to formulate a rigorous roadmap to the certification and qualification of AM metals and alloys is impeded by: (1) a lack of systematically quantified processing/structure/properties/performance (PSPP) data covering even a small

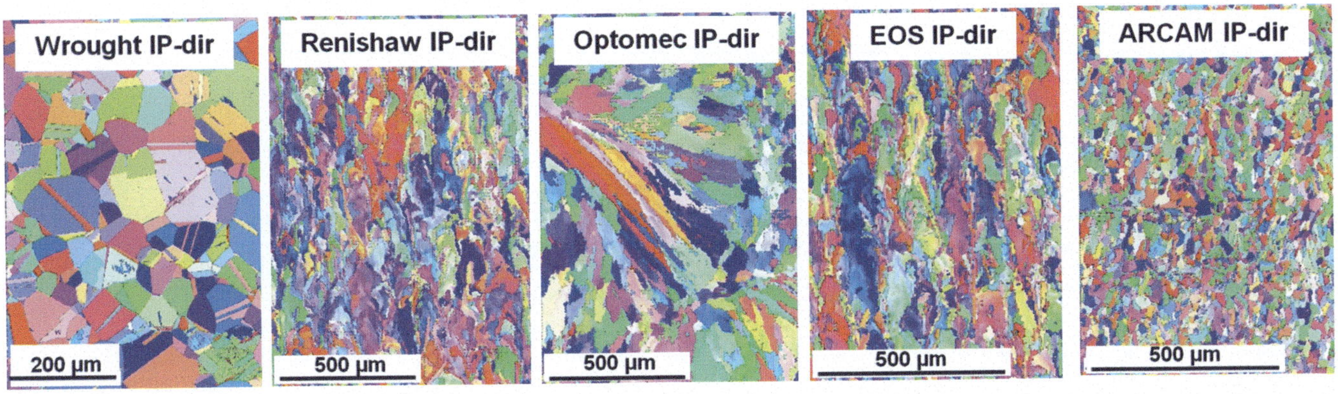

Fig. 1.1 EBSD images of the microstructures of 304 L SS: wrought and AM builds; In-plane direction produced on four different AM machines

fraction of the AM technologies/machines currently available [1], (2) few published archival research studies adequately reporting detailed specifics of feedstock pedigree (chemistry, structure, etc) linked to AM properties, (3) intellectual property concerns [7] limiting documentation/publication of detailed AM machine processing configurations and settings, and (4) quantified component/build thermal/ time histories, detailed build-orientation-mechanical testing properties, and detailed microstructure/texture and defect statistics [1–5].

References

1. Seifi, M., Salem, A., Beuth, J., Harrysson, O., Lewandowski, J.J.: Overview of materials qualification needs of metal additive manufacturing. JOM. **68**(**3**), 747 (2016)
2. Vaezi, M., Seitz, H., Yang, S.: A review on 3D MIcro-additive manufacturing technologies. Int. J. Adv. Manuf. Technol. **67**, 1721–1754 (2013)
3. Gray III, G.T., Livescu, V., Rigg, P.A., Trujillo, C.P., Cady, C.M., Chen, S.R., Carpenter, J.S., Lienert, T.J., Fensin, S.J.: Structure/property (constitutive and spallation response) of additively manufactured 316L stainless steel. Acta Mater. **138**, 140–149 (2017)
4. Frazier, W.E.: Metal additive manufacturing: a review. J. Mater. Eng. Perform. **23**(**6**), 1917 (2014)
5. Sames, W.J., List, F.A., Pannala, S., DeHoff, R.R., Babu, S.S.: The metallurgy and processing science of metal additive manufacturing. Int. Mater. Rev. **61**(**5**), 315 (2016)
6. Martin, J.H., Yahata, B.D., Hundley, J.M., Mayer, J.A., Schaedler, T.A., Pollock, T.M.: 3D printing of high-strength aluminum alloys. Nature. **549**, 365 (2017)
7. Osborn, L.S.: Reevaluating intellectual property law in a 3D printing era. Bridge. **47**(**3**), 18 (2017)

Chapter 2
Fatigue Characterization of 3D-Printed Maraging Steel by Infrared Thermography

Corentin Douellou, Xavier Balandraud, and Emmanuel Duc

Abstract Fatigue performances of additively manufactured metals are affected by the high temperature gradients and high cooling rates occurring during laser beam melting. More generally, mechanical properties strongly depend on the process parameters (scan speed, laser power, laser spot size, etc.). The present study proposes a calorific analysis of the fatigue response of 3D-printed maraging steel from thermal measurements obtained by infrared (IR) thermography. Fatigue damage is indeed associated with heat production leading to material self-heating. Analysis of the thermomechanical response was performed in two steps: first, measurement of temperature maps on the specimen surface by IR camera; second, calculation of the calorific origin of the temperature changes by image processing based on the heat diffusion equation. Using specific thermal data acquisition conditions, the processing enabled us to extract the heat power density corresponding to the mechanical dissipation caused by fatigue. The study was performed on specimens featuring a specific geometry and printed with different process parameters. Distinguishing differences in the production of mechanical dissipation at the beginning of fatigue tests can be useful to determine relevant configurations for long-term fatigue durability.

Keywords Fatigue · Self-heating · Infrared thermography · Laser beam melting · Maraging

2.1 Introduction

3D-printing, also known as "additive manufacturing" for industrial applications, is a quite novel manufacturing technology born in the 80's. Laser beam melting (LMB) technologies are of particular interest for cutting-edge industries (aeronautics, spatial, defense...) because it gives the possibility of complex geometry manufacturing of metallic parts. However industrial requirements in terms of mechanical properties are very severe while there is still a lack of knowledge on the physical mechanisms of the process. Several metals can be elaborated by LBM. Among them, maraging steels are well known for their high strength and toughness, and also for their numerous high performance industrial applications. Mechanical properties of additively manufactured metals are known to be very dependent on the process parameters. Indeed, these parameters influence the metallic microstructure and the occurrence of pores and residuals stresses. Nowadays, achievable tensile strength and hardness are comparable or even better than those of conventionally processed materials [1, 2]. Though, ductility and fatigue performances are strongly affected by the high temperature gradients and high cooling rates occurring during additive manufacturing process [1, 3, 4]. These general conclusions are also true for maraging steels [5, 6].

Classical fatigue limit estimations (S-N curve determination or staircase) are very time consuming and expensive. Generally one needs at least ten specimens for a staircase, and thirty specimens for a S-N curve determination. Each of these specimens is tested with a cyclic loading at constant amplitude until specimen failure. The fatigue limit corresponds to the amplitude for which a specimen is unlikely to fail at least before 10^7 cycles. In an industrial context for which testing cost and time are critical, it is extremely interesting to develop and apply rapid methods of fatigue limit estimation.

Any material produces or absorbs heat when it is submitted to mechanical loading. Thermoelastic couplings are involved in this phenomenon. Mechanical irreversibility is also accompanied by heat production. In particular, fatigue damage is associated with heat production leading to material self-heating. Analysis of the mechanical response can be performed by measuring temperature maps at the specimen surface by infrared (IR) thermography [7–9], and then deducing the calorific origin (heat sources, in $W.m^{-3}$) of the temperature changes [10–16]. The present study proposes a calorific analysis of the fatigue of a 3D-printed maraging steel, based on the use of a specific specimen geometry (see Fig. 2.1 and Ref. [17] for details).

C. Douellou · X. Balandraud (✉) · E. Duc
Université Clermont Auvergne, CNRS, SIGMA Clermont, Institut Pascal, Clermont-Ferrand, France
e-mail: xavier.balandraud@sigma-clermont.fr

S. Kramer et al. (eds.), *Mechanics of Additive and Advanced Manufacturing, Volume 8*, Conference Proceedings of the Society for Experimental Mechanics Series, https://doi.org/10.1007/978-3-319-95083-9_2

The paper is divided in three sections as follows. The first section provides the general background on the method used for heat source calculation. The second section provides details on the experimental set-up. Results are provided and discussed in the third section.

2.2 Methodology for Heat Source Calculation

Due to thermomechanical couplings, any material produces or absorbs heat when it is submitted to a mechanical loading. Heat sources are the corresponding heat power densities. Thermoelastic couplings lead to heat production or absorption as a function of the sign of the stress rate, whereas irreversible mechanical phenomena always lead to heat production. For example, fatigue damage is accompanied by a material self-heating. Based on the heat diffusion equation, heat sources can be reconstructed from the knowledge of the temperatures, captured for instance by IR thermography at the specimen surface. When heat sources are spatially homogeneous in the tested specimen, the reconstruction can be performed with a so-called 0D approach, which consists in using the mean temperature variation of the whole gauge zone (See refs. [13, 17, 18] for details). The simplified heat Eq. (2.1) provides the relation between the heat source $s(t)$ and the mean temperature change $\theta(t)$, where ρ is the material density (kg.m^{-3}), C the specific heat (J.kg^{-1}.°C^{-1}) and τ a time constant (s) characterizing the global heat exchanges with the outside of the specimen.

$$s(t) = \rho C\left(\frac{\mathrm{d}\theta(t)}{\mathrm{d}t} + \frac{\theta(t)}{\tau}\right) \tag{2.1}$$

Figure 2.1 presents the geometry of the plane specimen. It can be noticed that two reference elements are added on both sides of the mechanical tested part. Changes in specimen's environment during fatigue test can be tracked thanks to these reference elements: temperature of the jaws of the testing machine, temperature and flow of the ambient air, and more generally any change of the testing room conditions (machines, persons, walls, etc.).

Mechanical loading consisted of blocks of sinusoidal force-controlled cycles, with increasing amplitudes until specimen failure. In between each block, a stand-off period with no loading let the specimen return to ambient temperature. Figure 2.2 shows an example of the global temperature variation for one block, followed by the natural return to ambient temperature. For each loading block, the heat source can be then deduced from the global temperature θ(t) using Eq. (2.1). Mechanical dissipation due to fatigue damage can be extracted using a specific temperature acquisition frequency. Indeed, the heat produced by the thermoelastic coupling is null over a mechanical cycle. Considering the temperature change averaged over a mechanical cycle (or an integer number of cycles) thus directly provides the mechanical dissipation, see ref. [13] for more details. In practice, the temporal averaging operation can be performed in real time with certain IR cameras.

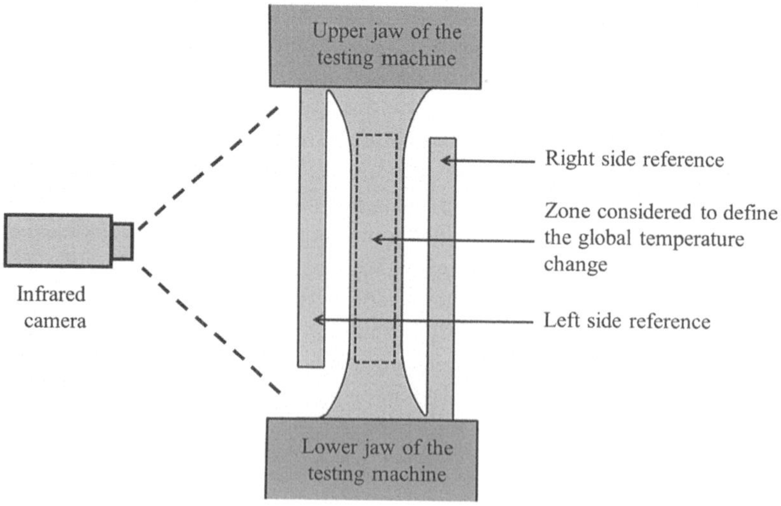

Fig. 2.1 Specimen geometry featuring two reference elements for the calorific analysis

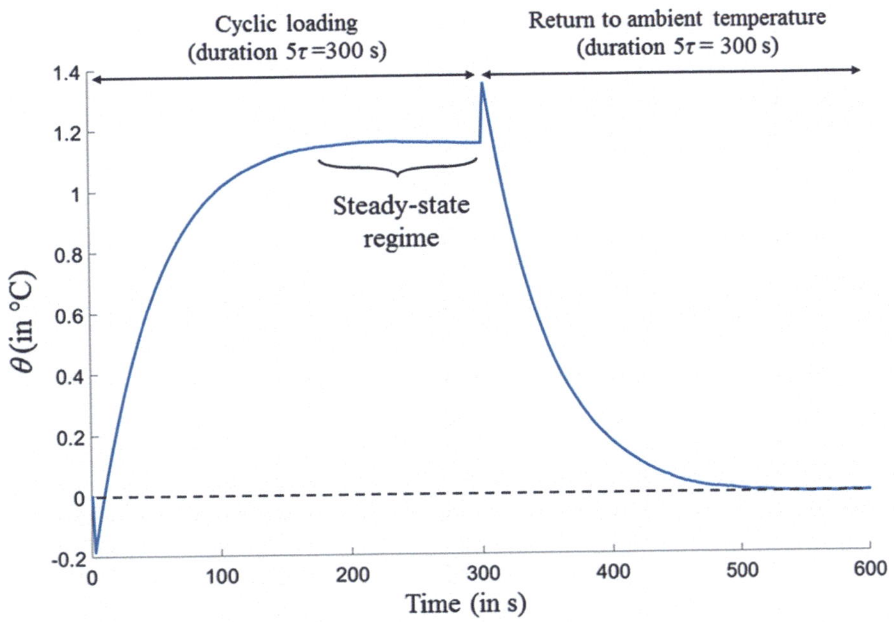

Fig. 2.2 Global temperature variation for one block of mechanical cycles followed by a return to ambient temperature

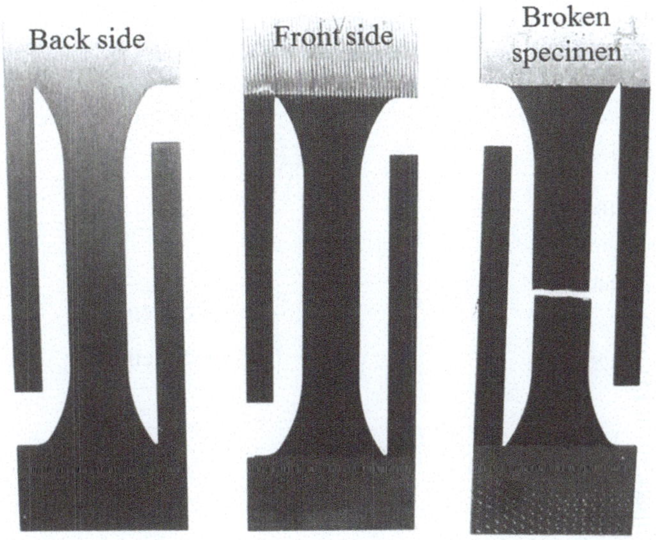

Fig. 2.3 3D-printed maraging specimens. The front side of each specimen was painted in black to maximize its thermal emissivity for temperature measurements during fatigue test

2.3 Experimental Set-Up

A Cedip Jade III-MWIR infrared camera was used to capture the temperature changes on the specimen surface. The loading was performed with a MTS uniaxial machine. Each specimen's front face was painted in black in order to maximize the thermal emissivity (set at 0.95), see Fig. 2.3. The specimen environment was also painted in black to limit parasitic reflections. Tests were performed on a maraging steel. Two types of specimen have been compared:

- Specimen 1: printed with high laser scan speed and large hatching distance in order to increase the machine productivity;
- Specimen 2: printed with a set of process parameters optimizing the material quality.

The same loading conditions were applied to both specimens: 300-s blocks of sinusoidal cycles with intermediate stand-off periods of the same duration. Stress amplitude was increased at each block of cycles, keeping the stress ratio and the frequency constant, until specimen failure. The natural returns to ambient temperature in between the blocks were used to calculate the value of the time constant τ involved in Eq. (2.1).

2.4 Analysis and Discussion

For both specimens, the mechanical dissipation *versus* stress amplitude plotting is presented in Fig. 2.4. Let us note first that specimen 1 broke earlier than specimen 2, respectively in the seventh and sixteenth blocks of cycles. Two regimes can be evidenced in the calorific response of the materials. For small stress amplitudes, mechanical dissipation increases slightly with the loading level. This behavior can be explained by grain reorientation [7]. A rapid increase in mechanical dissipation is observed for higher stress amplitudes. This type of plotting is usually employed in the literature by considering "temperature changes" (in °C) rather than "mechanical dissipation" (in kW/m^3) on the ordinate axis [8, 9]. It can be noted that the former quantity is not intrinsic to the material's fatigue response. Indeed, the temperature fields in the specimen depend on the heat diffusion, as well as on the heat exchange conditions at the boundaries (by contact with the jaws of the testing machine, and by convection with ambient air). Mechanical dissipation is a calorific quantity which depends only on the irreversible mechanical phenomena occurring in the material under test.

Similarly to what is done in the literature with temperatures, Fig. 2.4 can be employed to estimate the fatigue limit of the materials. It is possible to consider the crossing of the two asymptotes associated with the two regimes [8], or the intersection of the asymptote at the highest stress amplitudes with the abscissa axis [9]. It is interesting to compare the calorific responses of the two specimens. Fatigue limit of the specimen 2 is about twice higher than the one of specimen 1. It can be also noted that for a given loading level, mechanical dissipation is much greater for specimen 1 than for specimen 2. This means that the cyclic loading of specimen 1 is accompanied by much more fatigue damage in the same loading conditions. This last result was expected because of the differences in the two material elaboration processes. However it is worth noting that the conclusions were obtained by using one specimen of each type, for a test duration of a few hours.

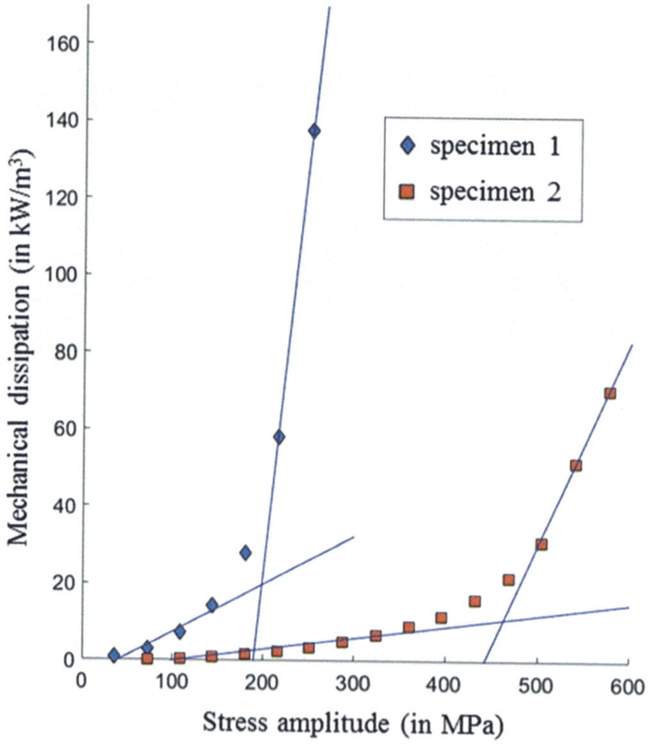

Fig. 2.4 Mechanical dissipation *versus* stress amplitude. Each point corresponds to a block of mechanical cycles lasting 300 s. Specimen 1 was printed with high laser scan speed and large hatching distance to increase the machine productivity. Specimen 2 was printed with a set of process parameters optimizing the material quality

2.5 Conclusion

This study shows that IR thermography can be employed to optimize the fatigue performances of maraging steels elaborated by additive manufacturing. The presentation will develop the reduction of parasitic effects caused by changes in the specimen's environment, the choice of relevant thermal acquisition parameters for mechanical dissipation calculation, and the signal processing. The impact of the process conditions on the fatigue performances will be discussed in terms of mechanical dissipation produced by the tested material. The use of this methodology for the manufacturing condition optimization is a prospect of the study. The methodology is indeed relevant to rapidly compare the fatigue performances of various elaboration parameters, which is particularly important for additive manufacturing where a constant compromise between productivity and mechanical performances has to be found.

Acknowledgements The authors gratefully acknowledge the Région Auvergne Rhône Alpes (*Ressourcement en fabrication additive*) for their financial aid, and the company AddUp, Cébazat, France, for their support during this research.

References

1. Riemer, A., Leuders, S., Thöne, M., Richard, H.A., Tröster, T., Niendorf, T.: On the fatigue crack growth behavior in 316L stainless steel manufactured by selective laser melting. Eng. Fract. Mech. **120**, 15–25 (2014)
2. Röttger, A., Geenen, K., Windmann, M., Binner, F., Theisen, W.: Comparison of microstructure and mechanical properties of 316 L austenitic steel processed by selective laser melting with hot-isostatic pressed and cast material. Mater. Sci. Eng. A. **678**, 365–376 (2016)
3. Rafi, H.K., Starr, T.L., Stucker, B.E.: A comparison of the tensile, fatigue, and fracture behavior of Ti-6Al-4V and 15-5 PH stainless steel parts made by selective laser melting. Int. J. Adv. Manuf. Technol. **69**, 1299–1309 (2013)
4. Spierings, A.B., Starr, T.L., Wegener, K.: Fatigue performance of additive manufactured metallic parts. Rapid Prototyp. J. **19**, 88–94 (2013)
5. Becker, T.H., Dimitrov, D.: The achievable mechanical properties of SLM produced Maraging steel 300 components. Rapid Prototyp. J. **22**, 487–494 (2016)
6. Suryawanshi, J., Prashanth, K.G., Ramamurty, U.: Tensile, fracture, and fatigue crack growth properties of a 3D printed maraging steel through selective laser melting. J. Alloys Compd. **725**, 355–364 (2017)
7. Munier, R., Doudard, C., Calloch, S., Weber, B.: Identification of the micro-plasticity mechanisms at the origin of self-heating under cyclic loading with low stress amplitude. Int. J. Fatigue. **103**, 122–135 (2017)
8. Luong, M.P.: Fatigue limit evaluation of metals using an infrared thermographic technique. Mech. Mater. **28**, 155–163 (1998)
9. La Rosa, G., Risitano, A.: Thermographic methodology for rapid determination of the fatigue limit of materials and mechanical components. Int. J. Fatigue. **22**, 65–73 (2000)
10. Chrysochoos, A., Peyroux, R.: Experimental analysis and numerical simulation of thermomechanical couplings in solid materials. Revue Générale de Thermique. **37**, 582–606 (1998)
11. Boulanger, T., Chrysochoos, A., Mabru, C., Galtier, A.: Calorimetric analysis of dissipative and thermoelastic effects associated with the fatigue behavior of steels. Int. J. Fatigue. **26**, 221–229 (2004)
12. Benaarbia, A., Chrysochoos, A., Robert, G.: Thermomechanical behavior of PA6.6 composites subjected to low cycle fatigue. Compos. Part B Eng. **76**, 52–64 (2015)
13. Delpueyo, D., Balandraud, X., Grédiac, M., Stanciu, S., Cimpoesu, N.: A specific device for enhanced measurement of mechanical dissipation in specimens subjected to long-term tensile tests in fatigue. Strain. **54**, e12252 (2018)
14. Le Cam, J.B., Samaca Martinez, J.R., Balandraud, X., Toussaint, E., Caillard, J.: Thermomechanical analysis of the singular behavior of rubber: entropic elasticity, reinforcement by fillers, strain-induced crystallization and the Mullins effect. Exp. Mech. **55**, 771–782 (2015)
15. Lachhab, A., Robin, E., Le Cam, J.B., Mortier, F., Tirel, Y., Canevet, E.: Thermomechanical analysis of polymeric foams subjected to cyclic loading: Anelasticity, self-heating and strain-induced crystallization. Polymer. **126**, 19–28 (2017)
16. Samaca Martinez, J.R., Le Cam, J.B., Balandraud, X., Toussaint, E., Caillard, J.: Mechanisms of deformation in crystallizable natural rubber. Part 1: Thermal characterization. Polymer. **54**, 2727–2736 (2013)
17. Jongchansitto P., Douellou C., Preechawuttipong I., Balandraud X., Comparison Between 0D and 1D Approaches for Mechanical Dissipation Measurement during Fatigue Tests, Strain, Submitted, 2018
18. Chrysochoos, A., Louche, H.: An infrared image processing to analyse the calorific effects accompanying strain localisation. Int. J. Eng. Sci. **38**, 1759–1788 (2000)

Chapter 3
Quasi-Static and Dynamic Fracture Behaviors of Additively Printed ABS Coupons Studied Using DIC: Role of Build Architecture and Loading Rate

John P. Isaac and Hareesh V. Tippur

Abstract This work deals with quasi-static and dynamic fracture behaviors of Additively Manufactured (AM) ABS coupons. In this research, planar specimens have been additively printed using a heated circular nozzle producing a continuous bead of 0.2 mm diameter and 0.2 mm layer thickness in the build directions. Of specific interest to this work is the role of two different print architectures namely 0/90° and ± 45° in-plane orientations. The local measurement of in-plane displacements are performed optically using DIC up to crack initiation and during growth in quasi-statically loaded 3-point bend specimens and impact loaded edge-notched specimens. The latter set of experiments are carried out using an instrumented Hopkinson pressure bar and an ultrahigh-speed digital camera. Significant differences in the engineering parameters and failure modes as a result of the build variables and loading rates are observed.

Keywords Digital image correlation · Additive manufacturing · Dynamic fracture · Stress intensity factors · Crack initiation

3.1 Introduction

Additive Manufacturing (AM) involves building a 3D solid layer-by-layer from a virtual model [1]. AM has restructured the industrial production process as complex geometries which are difficult to fabricate using conventional subtractive methods can be made more easily, efficiently and cost-effectively. The printing process, however, introduces artifacts which affect the mechanical behavior of the parts and hence need to be well understood. Currently there are different AM methods being used for various applications, the most popular one for thermoplastics being fused filament fabrication or FFF. In FFF, the molten polymer emerging out of a heated nozzle is laid down layer-by-layer in a predetermined computer-generated path on a heated platform. The option of prescribing the deposition path to build the object introduces anisotropy in terms of weak failure planes/paths. Though ABS - a common AM relevant polymer - is isotropic, the different paths traced by the deposition nozzle during AM produce macroscale orthotropic fracture characteristics resulting in mechanical and failure behavior differences.

In this work, 0/90° and ± 45° configurations were used to evaluate the tensile and fracture properties of additively produced ABS specimens. The optical method of Digital Image Correlation (DIC) was used to measure the local in-plane displacements in the whole field. As a vision-based optical technique, the DIC method has been widely accepted and commonly used as it is a powerful and flexible experimental mechanics tool for surface deformation measurement [2]. The undeformed and deformed images, recorded before and after the application of load, are correlated to find displacements in the entire region-of-interest. The measured displacements can be used in conjunction with the theoretical expressions and over-deterministic least-squares analyses to find the engineering parameters of interest [3]. In this work, tensile tests have been performed on dog-bone specimens to find quasi-static elastic constants and ultrasonic measurements to find the dynamic counterparts. Also, the crack tip Stress Intensity Factors (SIFs) have been calculated for quasi-statically loaded three-point bend specimens and impact loaded edge-notched specimens. The dynamic counterparts are currently being investigated.

J. P. Isaac · H. V. Tippur (✉)
Department of Mechanical Engineering, Auburn University, Auburn, AL, USA
e-mail: htippur@eng.auburn.edu

S. Kramer et al. (eds.), *Mechanics of Additive and Advanced Manufacturing, Volume 8*, Conference Proceedings of the Society for Experimental Mechanics Series, https://doi.org/10.1007/978-3-319-95083-9_3

3.2 Specimen Preparation

All AM specimens were made using a Cubicon 3DP-110F printer. The desired geometry of the specimen was created using Solidworks™ and saved as stereolithographic (STL) file. This STL file was then opened using Cubicreator, propriety slicer software available for the printer. Using this software, the necessary modifications with all the print settings were completed and the Gcode was exported to the printer. All the printing parameters except the direction of the build were same for both the configurations. These printing parameters are listed in Table 3.1. Also the infill rate was given as 100% in all the prints in order to avoid the effect of any additional porosity. The ABS required for printing was commercially purchased.

The two configurations used in this work are 0/90° and ± 45° in-plane orientations. In the 0/90° specimens, once the outer wall was built, the nozzle moved parallel to the *x*-axis (0°) applying the molten ABS along the prescribed path until the entire layer was built. Now moving onto the next layer, the nozzle moved perpendicular to the *y*-axis (90°) and the molten ABS was laid down accordingly. This process was repeated until the entire specimen was complete. Similarly, in the architecture of ±45° specimens, while building the first layer, the nozzle moved at an angle of +45° with the *x*-axis and then moved at −45° with the *x*-axis for the second layer, and so on.

3.3 Ultrasonic Measurements

Elastic constants measured under dynamic conditions are appropriate when evaluating the SIFs from displacement field equations for impact loading conditions. Olympus model Epoch 600 ultrasonic tester was used to find these constants. A 12.7 mm cube specimen was printed in both the 0/90° and ± 45° configurations. Since the elastic constants could vary in any of the thickness-wise directions, specimens were examined along all the three axes-of-rotation. The various values calculated are listed in the Table 3.2. Time-of-flight for longitudinal and shear waves to travel the thickness in each particular direction was measured using longitudinal and shear transducers, respectively. The equations used to calculate the dynamic Poisson's ratios and elastic moduli are given below:

$$C_L = \sqrt{\frac{E(1-\nu)}{\rho(1+\nu)(1-2\nu)}}, C_S = \sqrt{\frac{E}{2\rho(1+\nu)}} \tag{3.1}$$

Based on the measured data, the variation in the elastic properties are negligible in either of the print architectures which in turn indicates that the printed material can be treated as elastically isotropic solid.

Table 3.1 AM parameters used to print ABS specimens

Printing parameters	Values	Printing parameters	Values
Extruder temperature	240 °C	Layer thickness	0.2 mm
Bed temperature	115 °C	Wall thickness	0.8 mm
Chamber temperature	50 °C	Infill speed	80 mm/s

Table 3.2 Ultrasonic measurements

Print architecture	Axis of measurement	Longitudinal velocity (m/s)	Shear velocity (m/s)	Density (Kg/m3)	Poisson's ratio	Elastic modulus (GPa)
±45°	X	2041	984	1020	0.349	2.67
±45°	Y	2038	986	1020	0.347	2.67
±45°	Z	2061	988	1020	0.351	2.69
0/90°	X	2045	987	1010	0.348	2.66
0/90°	Y	2041	986	1010	0.348	2.65
0/90°	Z	2035	983	1010	0.348	2.63

3.4 Experimental Setup

3.4.1 *Quasi-Static Tension & Fracture Tests*

Dog-bone specimens of thickness 4 mm were printed for both 0/90° and ± 45° configurations according to the ASTM D638-14 standard. A fine coat of random speckles was sprayed on one of the surfaces of the specimen to measure deformations using DIC. Instron 4465 mechanical tester fitted with a 5 kN load-cell was used for these tensile tests at a crosshead speed of 0.05 mm/s. Load, displacement and strain data were all collected at 0.1 s intervals. A Pointgrey digital camera captured the speckle images once every 2 s. By correlating the images in the deformed state with the one in the reference state, two orthogonal in-plane displacement fields were obtained at each time instant or load-step. The specimens of both architectures were tested until fracture to get the entire stress-strain response.

Figure 3.1 shows dimensions of the three-point bend specimen used for quasi-static fracture tests. In this specimen, a crack of 6 mm was inserted using a diamond impregnated circular saw and the crack was sharpened prior to the test using a sharp razor blade. Instron 4465 machine was used to perform the quasi-static fracture experiments at a crosshead speed of 0.007 mm/s using a 5 kN load cell. Load and displacement data were collected every 0.1 s intervals. As in the tensile tests, speckles were coated on one of the surfaces of the specimen to perform displacement measurements using DIC. A Pointgrey camera (max. resolution 2048 × 2048 pixels) was used to record images once every second. Specimens were printed in both the printing configurations and tested until fracture. The undeformed and deformed images until crack initiation was correlated to find the crack initiation SIFs.

3.4.2 *Dynamic Fracture Tests*

Figure 3.2 shows the geometry of the V-notched specimens. A crack of length 3 mm was inserted into the specimen at the base of the V-notch tip and then sharpened by a razor blade. Ultrahigh speed digital photography was employed to record random speckles sprayed on to the specimen surface and perform DIC subsequently to study deformations during fracture of the two

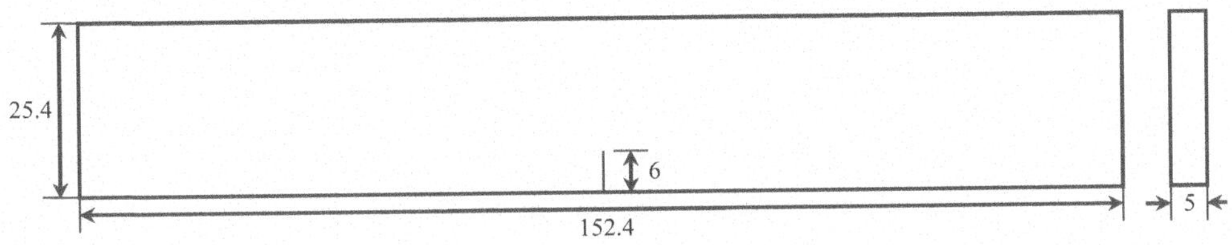

Fig. 3.1 Three-point bend quasi-static fracture test specimen (all dimensions are in mm)

Fig. 3.2 Dynamic fracture specimen (all dimensions are in mm)

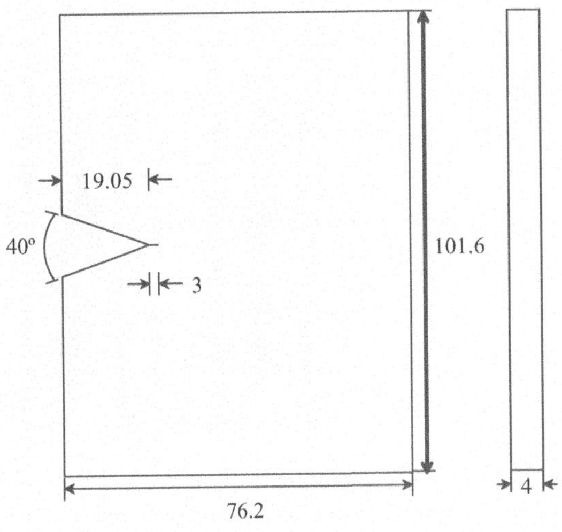

AM specimen architectures. A Hopkinson pressure bar (long-bar) was used for loading the notched specimen. The length of the long-bar was 1.83 m and diameter was 25.4 mm. One end of the long-bar had a smooth surface with which the striker rod made contact when propelled from the barrel of a gas-gun. The other end of the long-bar was shaped as a wedge to be inserted into the V-notch in the specimen. The striker rod was of length 305 mm and diameter 25.4 mm and it was held in the barrel of a gas-gun aligned co-axially with the long-bar. When the gas chamber was depressurized suddenly, the striker rod moved at a high velocity and made contact with the flat end of the long-bar which in turn delivered the impact to the specimen. Both the long-bar and striker rod used were made of AL 7075-T6 to avoid any impedance mismatch. The striker velocity of 20 ± 0.5 m/s was used in all the tests performed.

3.5 Results

3.5.1 Tensile Tests

The stress-strain response for the two print architectures of the same geometry were studied. The resulting data is summarized in Table 3.3. Evidently, there was a significant increase in the load (~60%) at which failure occurred for the ±45° specimen relative to the 0/90° counterpart. The corresponding specimen extension and failure strain is also higher (~32%) for the ±45° suggesting a higher ductility and toughness. Table 3.3 also shows the values of the elastic and failure properties relative to the ones found in the literature for cast ABS. Figure 3.3 shows the optical micrographs of the fractured surfaces for the two architectures. The micrographs (*x-z* plane) show striking differences responsible for the macroscale property variations listed in Table 3.3. Evidently, the 0/90° architecture shows substantial porosity aligned in one of the directions and shows significant thinning of the layers during the failure process leaving behind well-aligned and continuous gaps between the disbonded layers responsible for the lower failure stress and strains. Comparatively, the ±45° architecture also shows significant porosity but with a staggered alignment of pores. The micrographs also show higher tortuosity and stepped/staircase pattern on the

Table 3.3 Material properties obtained from tension tests

Material property	0/90° specimen	±45° specimen	Cast
Elastic modulus (GPa)	2.16 ± 0.20	2.34 ± 0.20	2.25[a]
Poisson's ratio	0.34 ± 0.10	0.34 ± 0.15	–
Failure load (N)	1550	2484	–
Failure stress (MPa)	23.4	37.4	42.5
Failure strain %	1.38	1.82	24

[a]Flexural modulus

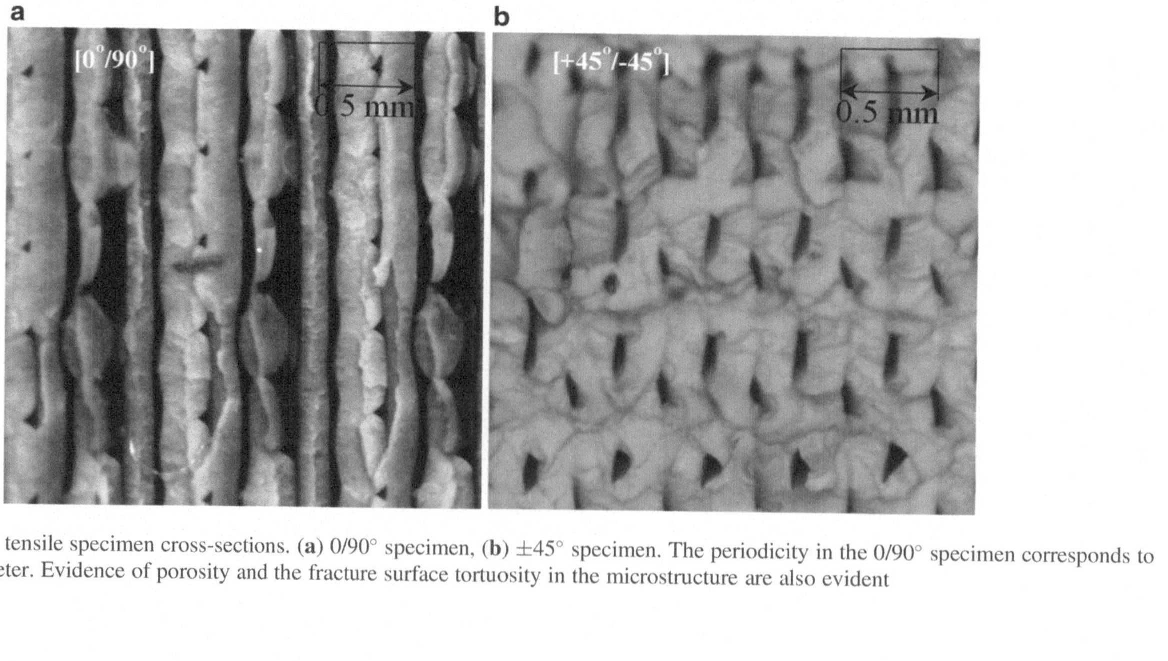

Fig. 3.3 Fractured tensile specimen cross-sections. (**a**) 0/90° specimen, (**b**) ±45° specimen. The periodicity in the 0/90° specimen corresponds to 0.2 mm bead diameter. Evidence of porosity and the fracture surface tortuosity in the microstructure are also evident

failed ligaments suggestive of shear band formation during failure. These are attributed to the relatively higher failure strength of this architecture.

3.5.2 *Quasi-Static Fracture Tests*

The load vs load-point deflection data for edge cracked 3-point bend specimens of identical geometry are plotted in Fig. 3.4 for the two print architectures. As in the tensile tests, significantly higher (almost by a factor of two) crack initiation load and hence the crack initiation toughness for the ±45° case relative to the 0/90° is evident. Furthermore, the crack growth occurred relatively rapidly and in a brittle fashion, as indicated by the quick drop in the load value beyond the peak load at crack initiation, in the 0/90° specimen. On the contrary, the crack initiation and growth in the ±45° specimen was different with a zig-zag/staircase growth pattern along a macroscopically ±45° bias after crack initiation. The drop in the load was gradual with a significant monotonic increase in the load-point displacement due to ductility offered by the print architecture. The speckle images captured by the camera in the deformed state were correlated with the reference images taken before the application of load to obtain the displacement fields. The speckle images were correlated using ARAMIS image analysis software. Using displacements in conjunction with an over-deterministic least-squares analysis, SIFs at each time step or load level were extracted [4]. Figures 3.5 and 3.6 show one of the captured images used for correlation for the 0/90° and ± 45° configurations, respectively. It can be seen from the images that the path of crack propagation is very much different in both the specimens. In the 0/90° specimen, the crack propagates along a nominally 0° path whereas in its counterpart, the crack is

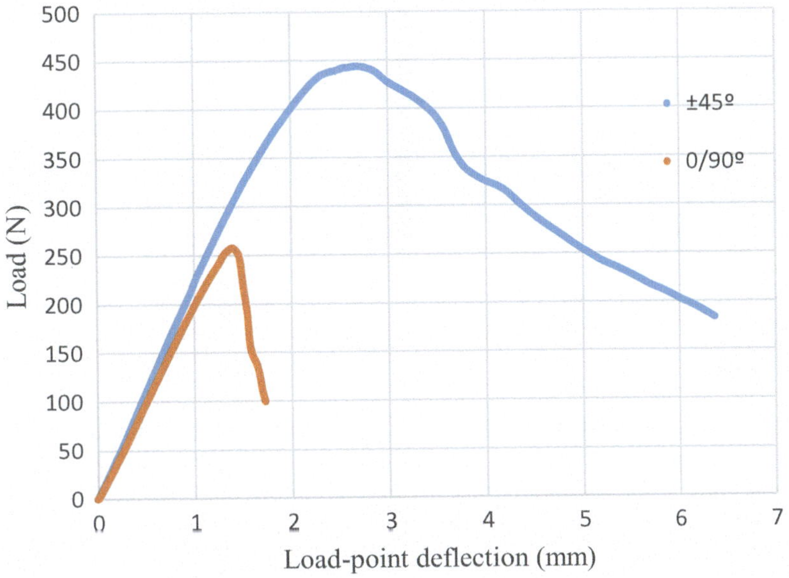

Fig. 3.4 Load vs Load-point displacement plots for quasi-static loading of edge cracked 3-point bend specimens

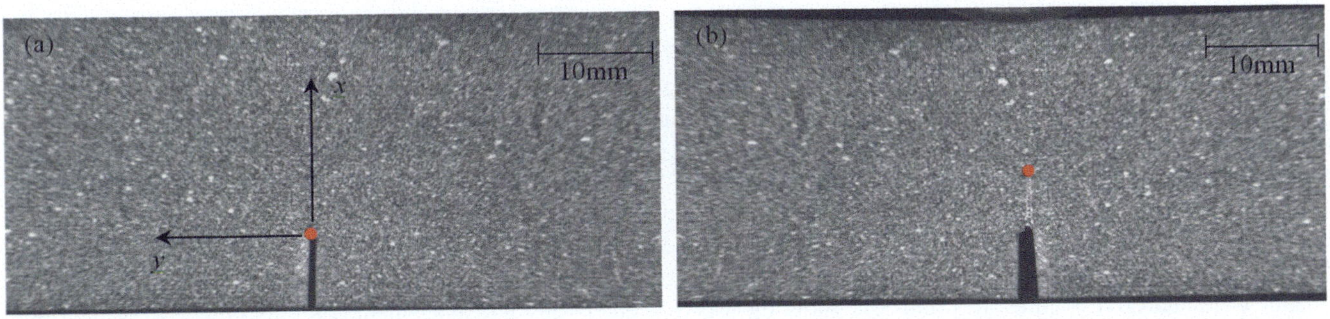

Fig. 3.5 Deformed images of 0/90° specimen under quasi-static loading (undeformed crack length = 6 mm) (**a**) Before crack initiation. (**b**) After crack initiation. Red dot indicates the crack tip at this time step

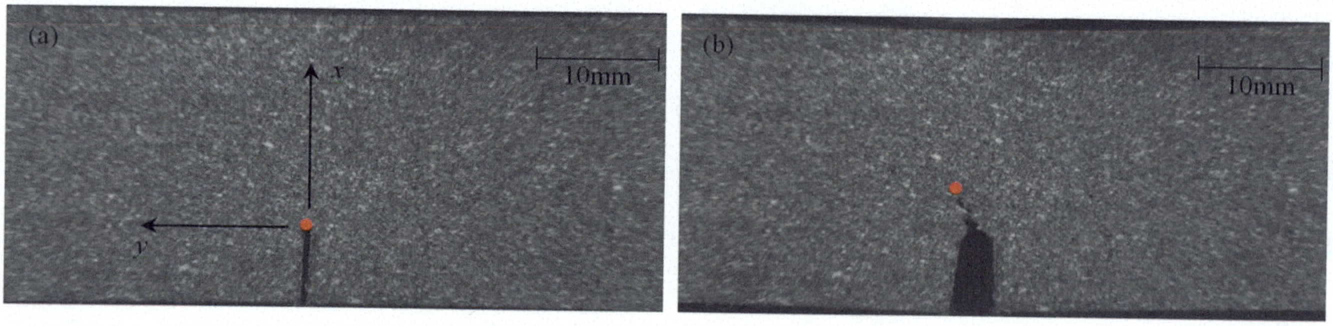

Fig. 3.6 Deformed images of ±45° specimen under quasi-static loading (undeformed crack length = 6 mm) (**a**) Before crack initiation. (**b**) After crack initiation. Red dot indicates the crack tip at this time/load step

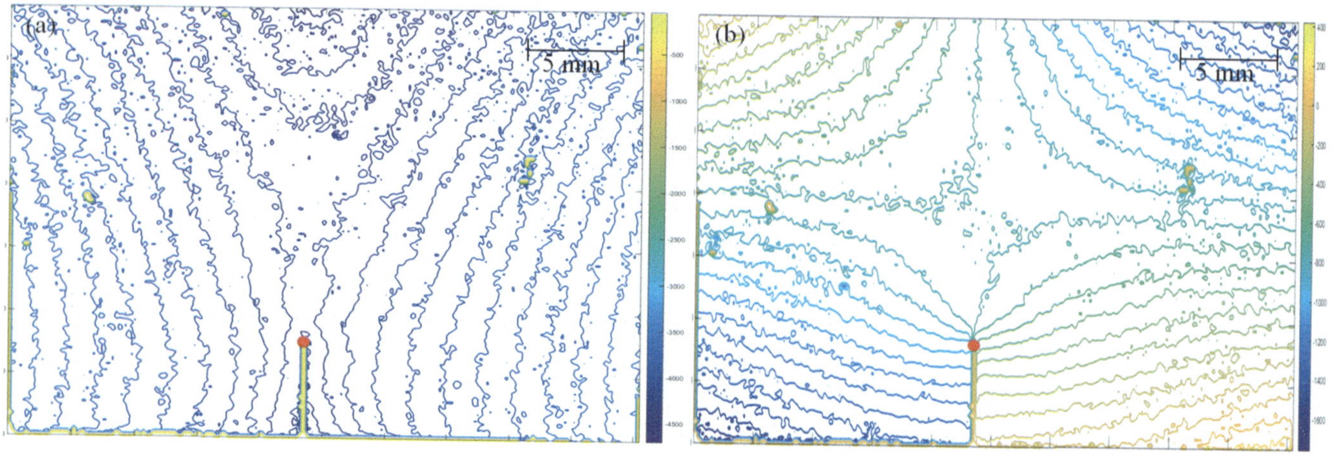

Fig. 3.7 Displacements contours of 0/90° configuration, for a load of 208 N with contour interval of 75 μm, obtained from DIC (undeformed crack length = 6 mm) (**a**) *x*-displacement. (**b**) *y*-displacement. Red dot indicates the crack tip at this time step and color-bar indicates displacement in micrometers

along a 45° path with a distinct staircase pattern. Furthermore, significant crack tip blunting occurred in both the cases with noticeably higher amount in the ±45° case. Figures 3.7 and 3.8 show samples of the *x*-displacement and *y*-displacement contours obtained for the 0/90° and ± 45° configurations, respectively, from DIC. Based on the different contour increments used in each contour plots, the 0/90° specimen has undergone substantially higher deformation for the same load relative to the ±45° counterpart.

3.5.3 Dynamic Fracture Tests

The V-notched specimens were dynamically loaded by the wedge-tipped long-bar by launching a striker towards it at a velocity of ~20 m/s. The stress waves transmitted into the specimen loaded the sharpened notch tip to initiate a crack along the 0° and ± 45°crack paths in the two architectures. The specimen surfaces were coated with random speckles and were photographed using Kirana 05 ultrahigh-speed single sensor camera (framing rate 200,000 fps) capable of recording 180, 10-bit, 924 × 768 pixel images. The images in the deformed state were correlated with the one in the undeformed state recorded before the start of the impact event [5]. Figures 3.9 and 3.10 show one of the captured images used for correlation for the 0/90° and ± 45° configurations respectively. The difference in the path of crack propagation was clearly visible when the images were analyzed. Figures 3.11 and 3.12 show the *x*-displacement and *y*-displacement contours obtained for the 0/90° and ± 45° configurations respectively. In the 0/90° configuration, the crack initiates along 0° and propagates in the same path as a mode-I crack until it reaches the opposite edge of the specimen. In the ±45° configuration, however, after crack initiation

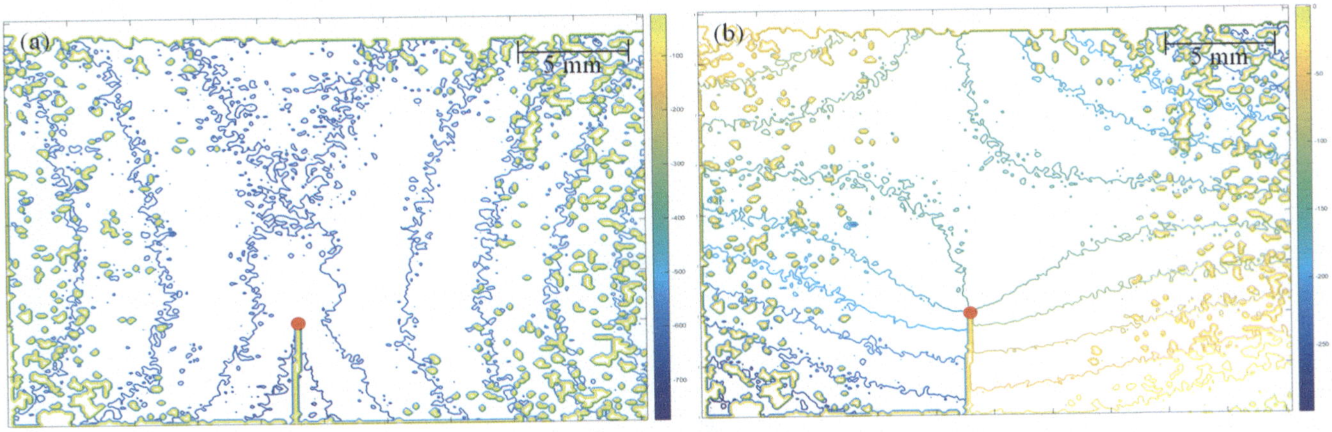

Fig. 3.8 Displacements contours of ±45° configuration, for a load of 208 N with contour interval of 25 μm, obtained through DIC (undeformed crack length = 6 mm) (**a**) x-displacement. (**b**) y-displacement. Red dot indicates the crack tip at this time step and color-bar indicates displacement in micrometers

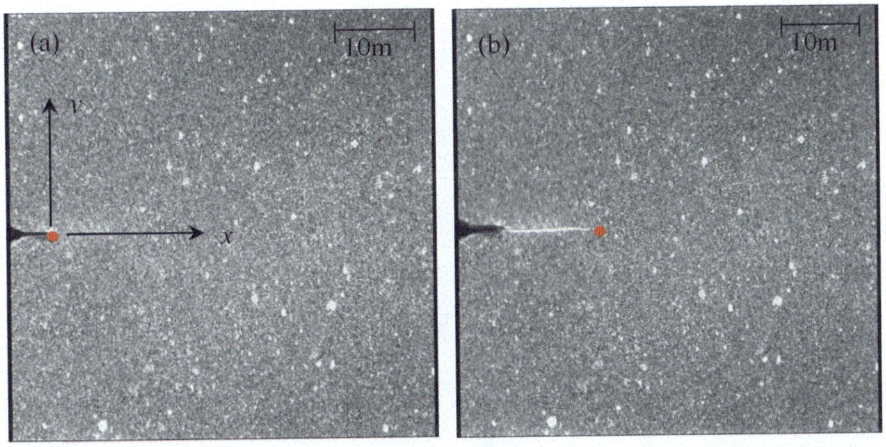

Fig. 3.9 Deformed images of 0/90° specimen under dynamic loading. (**a**) Before crack initiation. (**b**) After crack initiation. Red dot indicates the crack tip at this time step

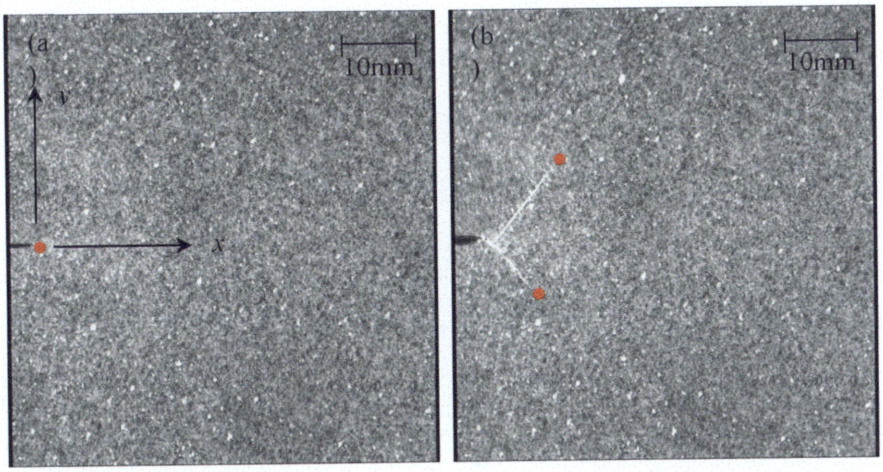

Fig. 3.10 Deformed images of ±45° specimen under dynamic loading. (**a**) Before crack initiation. (**b**) After crack initiation. Red dot indicates the crack tip at this time step

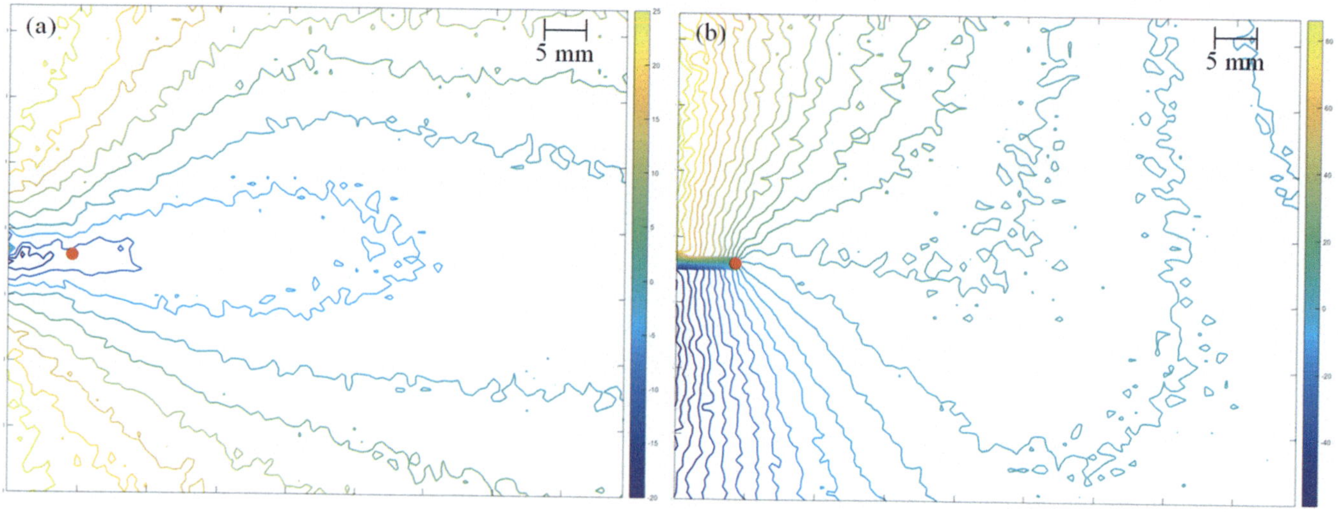

Fig. 3.11 Displacements contours of 0/90°, with contour interval of 5 μm, obtained through DIC (**a**) x-displacement. (**b**) y-displacement. Red dot indicates the crack tip at this time step

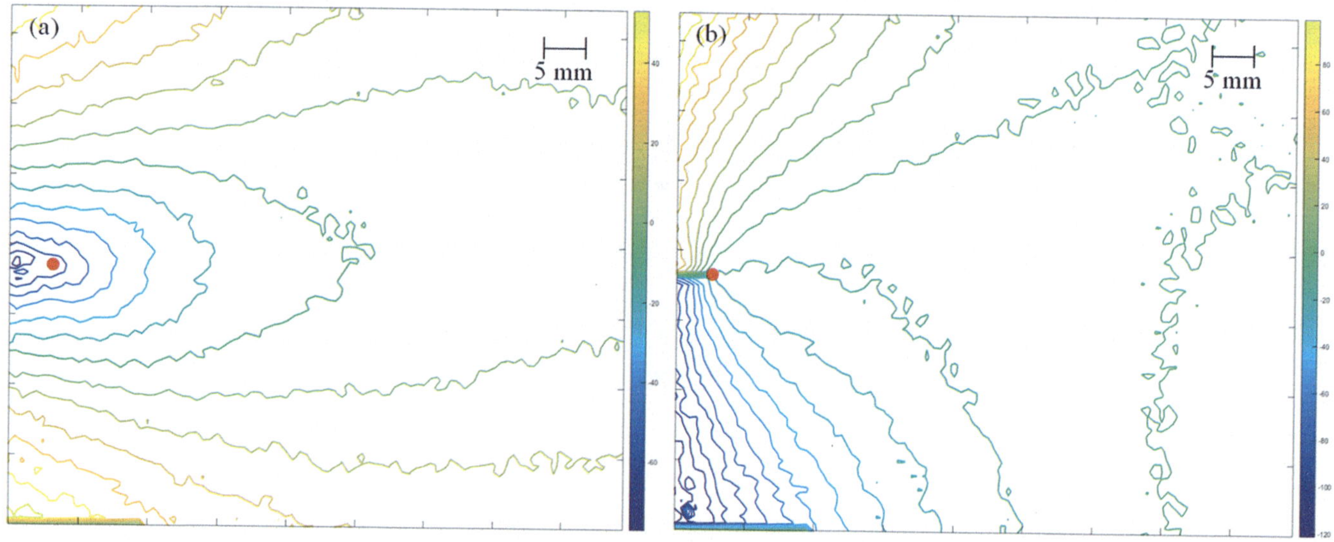

Fig. 3.12 Displacements contours of ±45°, with contour interval of 10 μm, obtained through DIC (**a**) displacements in the x-direction. (**b**) displacements in the y-direction. Red dot indicates the crack tip at this time step

the crack propagated along −45° for a small distance during which a new crack branch was generated and propagated along +45° path.

This new crack propagated successfully until the the specimen successfully whereas the initial crack (−45° path) died out.

3.6 Summary and Work-In-Progress

The quasi-static and dynamic fracture behaviors of additively printed ABS are studied in order to understand the role of print architecture and the loading rate of AM parts. The optical method of 2D-DIC is employed to map crack tip deformations in the whole field. The dynamic experiments are performed by utilizing DIC in conjunction with ultrahigh-speed photography and modified Hopkinson pressure bar loading facility. Distinctly different failure loads, fracture surface morphologies and crack growth patterns are evident in the two print architectures, 0/90° and ± 45°, studied. The optically measured displacement fields are currently being analyzed to extract the crack length, crack velocity and SIF histories during the quasi-static and dynamic loading events to assess fracture resistance behaviors. The measurements will be reconciled with the fracture surface morphology to assess the process-microstructure-property relationship.

References

1. Standard Terminology for Additive Manufacturing Technologies. ASTM International Designation: F2792-12a (2012)
2. Pan, B., et al.: Two-dimensional digital image correlation for in-plane displacement and strain measurement: a review. Meas. Sci. Technol. **20**(6), 062001 (2009). https://doi.org/10.1088/0957-0233/20/6/062001
3. Yoneyama, S., et al.: Evaluating mixed-mode stress intensity factors from full-field displacement fields obtained by optical methods. Eng. Fract. Mech. **74**(9), 1399–1412 (2007). https://doi.org/10.1016/j.engfracmech.2006.08.004
4. Lee, D., et al.: Experimental study of dynamic crack growth in unidirectional graphite/epoxy composites using digital image correlation method and high-speed photography. J. Compos. Mater. **43**(19), 2081–2108 (2009)
5. Kirugulige, M.S., et al.: Measurement of fracture parameters for a mixed-mode crack driven by stress waves using image correlation technique and high-speed digital photography. Strain. **45**, 108–122 (2009). https://doi.org/10.1111/j.1475-1305.2008.00449.x

Chapter 4
Compression and Shear Response of 3D Printed Foam Pads

Wei-Yang Lu

Abstract Polymeric porous materials have a wide range of applications. An important one in structural engineering is to use foams for cushioning or absorbing the kinetic energy from impact. Conventional foaming processes produce polymeric foams with disordered three-dimensional networks, which are dispersion in cell shape, size, etc. Since mechanical properties depend on the shape and structure of the cell, these foams are difficult to characterize and predict due to complexity and variation of cells. The new 3D printing fabrication method can now prepare components of foams with perfect regular array of cells. The printed foams potentially could be tuned or designed for application. In this study, foam pads of various porosities were printed using the same polymer. They all have a Body Centered Cubic (BCC) cell structured but with different span sizes. Experiments were conducted to characterize these foam pads in compression and shear, including off-axis loadings. The property of printing polymer was also characterized for analyzing the behaviors of these foam pads. Results are compared.

Keywords Additive manufacture · Silicone foam · Compression pad · Shear characterization

4.1 Introduction

In 3D printing polymers, such as silicone, can be heated and deposited layer by layer under computer control to build 3D foam objects. Instead of random cell shape and size, the 3D printed foams can have perfect array of cells. The mechanical properties of a printed foam would be easier to predict if the cell parameters are known. This study concentrates on characterizing the mechanical behaviors of foams with BCC cell structure. The BCC cell foam is defined by two printing parameters: span size (SS) and filament diameter (FD) as shown in Fig. 4.1. For various foam pads considered here, the constitutive polymer was SE1700 and FD was kept constant at 0.25 mm; SS was the only variable, its value ranged from 0 to 2.0 mm. The figure shows the foam pad with SS = 1.50 mm.

4.2 Compression Characterization

Punch compression test was performed as shown in Fig. 4.2a, where the platen was smaller than the pad sample. The platen diameter was about 31.56 mm; the nominal pad size was 50 mm × 50 mm × 3 mm. The loading was quasi-static, about 0.001 s^{-1}. The normalized load-displacement curves are plotted in Fig. 4.2b–d; (b) and (c) are the same plot but with different scales, so the responses of lower density pads are distinguishable; (d) displays the loading-unloading hysteresis. In addition to printed foam pads, solid SE1700 and a conventional silicone foam are also included in the plots for comparison. Notice that the pad that SS = 0 mm and solid SE1700 are not the same.

W.-Y. Lu (✉)
Sandia National Laboratories, Livermore, CA, USA
e-mail: wlu@sandia.gov

S. Kramer et al. (eds.), *Mechanics of Additive and Advanced Manufacturing, Volume 8*, Conference Proceedings of the Society for Experimental Mechanics Series, https://doi.org/10.1007/978-3-319-95083-9_4

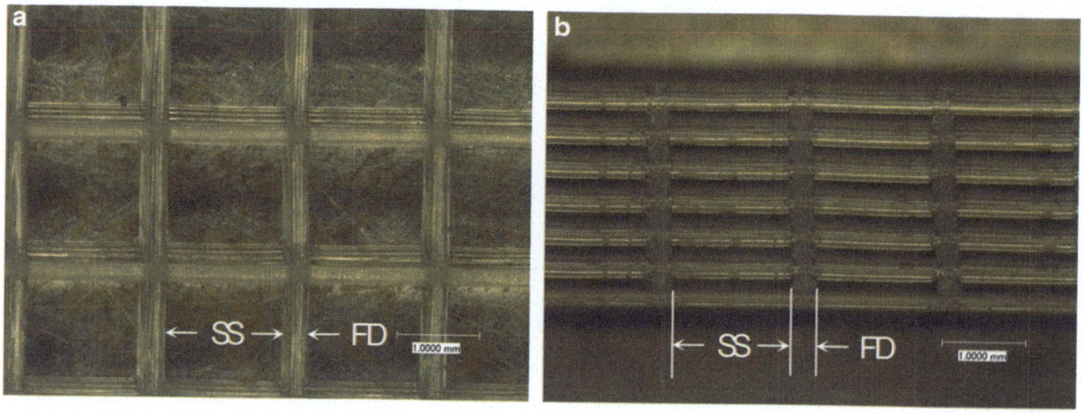

Fig. 4.1 3D printed foam pad with BCC cell structure. (**a**) Top view (**b**) Side view

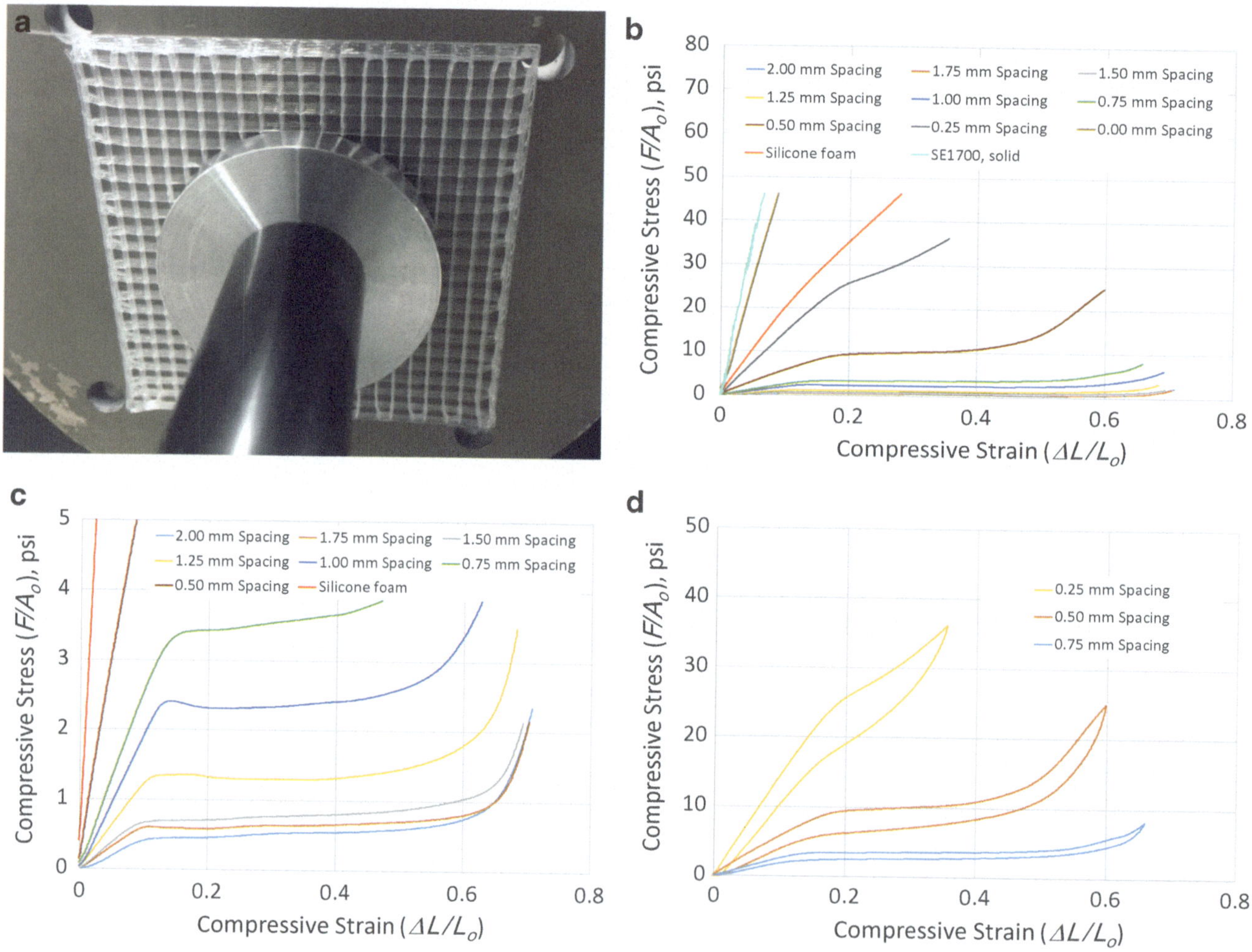

Fig. 4.2 Compression of 3D printed foam pad. (**a**) Compression setup (**b**) $\sigma - \varepsilon$ curves (**c**) $\sigma - \varepsilon$ curves (**d**) Hysteresis curves

4.3 Shear Characterization

Printed foam pad was cut to about 18 mm × 18 mm × 3 mm size and bonded to plates for lap shear test. Transparent plates, such as glass or Plexiglas, were used, so the integrity of bonding interfaces and the uniformity of foam cell deformation could be observed and assured. Since the pad was orthotropic, the shear was done in two orientations: one was the filament either parallel or perpendicular to the shear loading direction; the other was off-axis, which had an angle of 45° between filament and loading directions. These configurations were termed as 90 and 45, respectively, shown in Fig. 4.3. The figure also shows top views of un-deformed and shear deformed foam specimens, where SS = 1.75 mm. On shear deformed specimens, Fig. 4.3b, d, filaments from lower layers become visible.

Shear stress-strain curves are shown in Fig. 4.4. Figure 4.4b indicates that configuration 45 is softer than 90. Other pads, i.e. those with different SS values, also show the same trend.

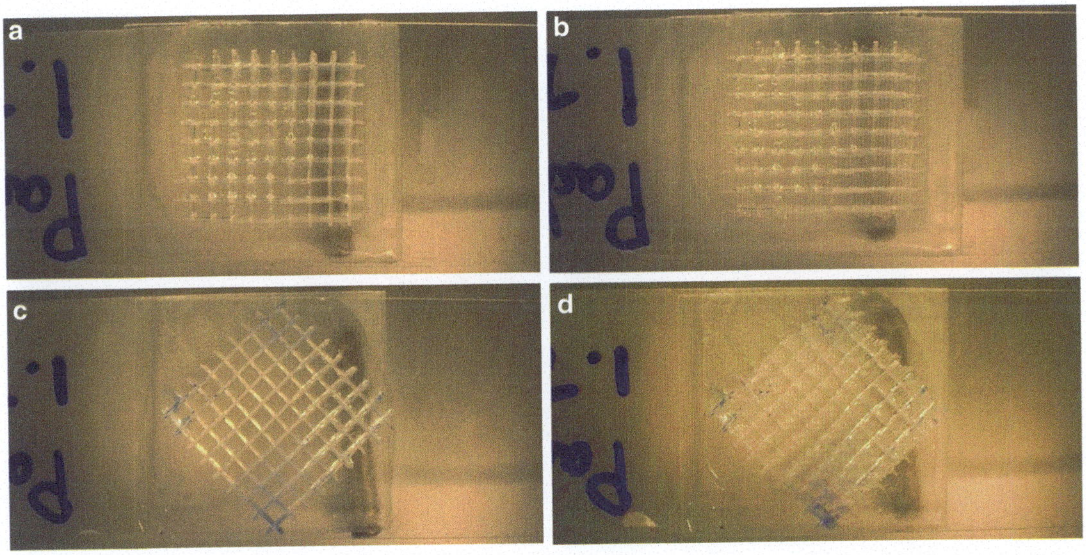

Fig. 4.3 Shear specimens and deformation, SS = 1.75 mm. (**a**) 90, un-deformed (**b**) 90, deformed (**c**) 45, un-deformed (**d**) 45, deformed

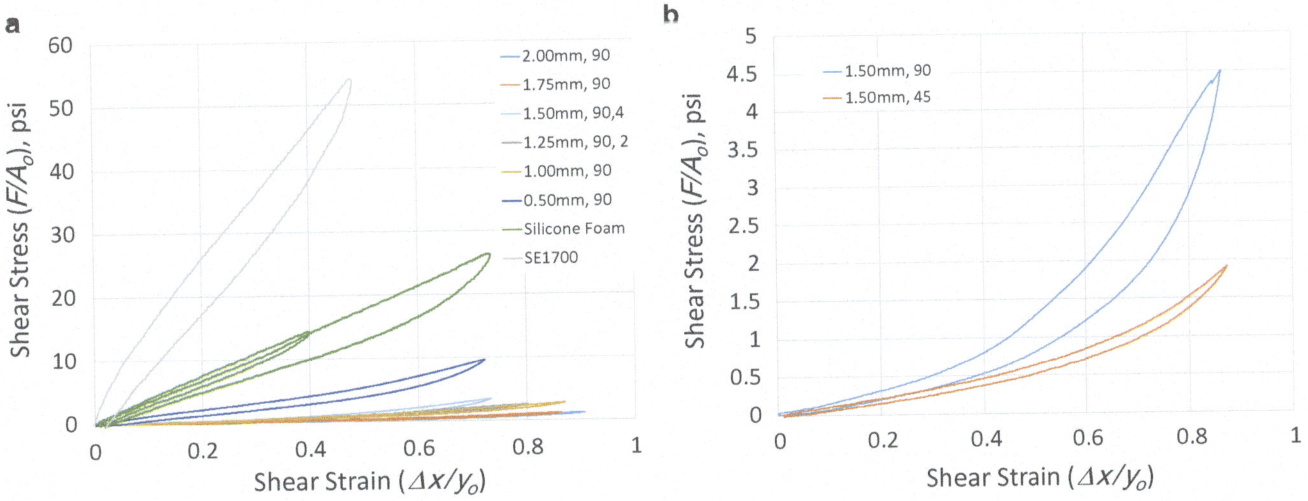

Fig. 4.4 Shear stress-strain curves. (**a**) 90, $\tau - \gamma$ curves (**b**) 90 and 45 comparison

4.4 Summary and Conclusion

The compression and shear properties of various 3D printed foam pads were characterized. Only considering one cell shape and one cell variable, span size, a wide range of mechanical properties of printed foam pads, from compliant to stiff, can already be achieved by additive manufacturing process.

Acknowledgements Thank Adam Cook, Robert Bernstein, and Alexis Abelow for providing materials and discussions. Sandia National Laboratories is a multimission laboratory managed and operated by National Technology and Engineering Solutions of Sandia, LLC, a wholly owned subsidiary of Honeywell International, Inc., for the U.S. Department of Energy's National Nuclear Security Administration under contract DE-NA0003525.

Chapter 5
Mechanical Structure-Property Relationships for 2D Polymers Comprised of Nodes and Bridge Units

Emil Sandoz-Rosado and Eric D. Wetzel

Abstract 2D polymers have emerged as an infinitely-tailorable material with remarkable, tunable response and density-normalized mechanical properties far exceeding structural materials such as steel, high-performance fibers or reinforced composites. It is critical that the vast material design space of 2D polymers be mapped in order to achieve optimal mechanical performance, since hundreds of permutations of one class of 2D polymers known as covalent organic frameworks have already been synthesized in the decade since the introduction of these materials. To this end, this work establishes a general structure-property relationship for elastic modulus and strength for a common 2D polymer motif consisting of nodes linked by linear bridge polymer chains to form a two-dimensional network. The length of the bridge chains are parametrically varied to study the impact of chain compliance on stiffness and strength. The density-normalized isotropic strength of the graphene/polyethylene hybrid material known as graphylene begins at 0.015 GPa/kg·m^3 (50% higher than that of perfect crystalline Kevlar®) and the density-normalized isotropic stiffness is 0.143 GPa/kg·m^3 (31% higher than Kevlar®) and decreases non-monotonically with increasing bridge chain length. The mechanical response is mapped and correlated to the inherent molecular structure of these general 2D polymer as a framework for designing 2D polymer molecules for mechanical applications from the ground up.

Keywords 2D polymer · Mechanics · Fracture · Strength · Design

5.1 Results and Discussion

Many 2D polymers, especially those known as covalent organic frameworks, can be described as a 2D network of nodes interconnected by linear linkages or bridge units. One of the simplest representations of this node-bridge network is a repeating motif of a benzene ring connected by six polyethylene chains at each carbon in the aromatic ring, which we have called 'graphylene', or GrE-n where 'n' denotes the number of ethylene mers in each bridge unit chain (e.g. GrE-4 denotes graphylene with 4 mers of polyethylene in each of the six bridge unit chains). In this work, the bridge unit length for graphylene is parametrically varied to examine the impact of bridge chain length on mechanical properties as predicted by atomistic simulations. GrE-n unit cells were determined through density functional theory, and were then simulated as individual monolayers using molecular dynamics to determine stiffness and strength in a larger domain (Fig. 5.1a) for GrE-2 through GrE-6 (GrE-2 depicted in Fig. 5.1b and GrE-5 depicted in Fig. 5.1c). The objective is to gain insights as to the desired molecular arrangements, and establish structure-property relationships for achieving materials with superior mechanical properties through intentional design.

The AIREBO potential was used for these simulations with an appropriate cutoff distance of 0.197 nm. To determine the mechanical properties, the domains were strained perpendicular to the GrE-n first nearest neighbor (1NN) crystalline direction, with the 1NN direction defined as beginning from one nodal benzene ring and following the bridge chain to the next nearest nodal benzene ring. The domain was equilibrated to 0 K and then time integrated with constant energy (NVE) with a time step of 0.5 fs and deformed and a constant strain rate of 2 ns^{-1}. Mechanical properties are reported in terms of engineering stress and strain, and in the two-dimensional formulation.

E. Sandoz-Rosado · E. D. Wetzel (✉)
US Army Research Laboratory, Composite & Hybrid Materials Branch, Adelphi, MD, USA

S. Kramer et al. (eds.), *Mechanics of Additive and Advanced Manufacturing, Volume 8*, Conference Proceedings of the Society for Experimental Mechanics Series, https://doi.org/10.1007/978-3-319-95083-9_5

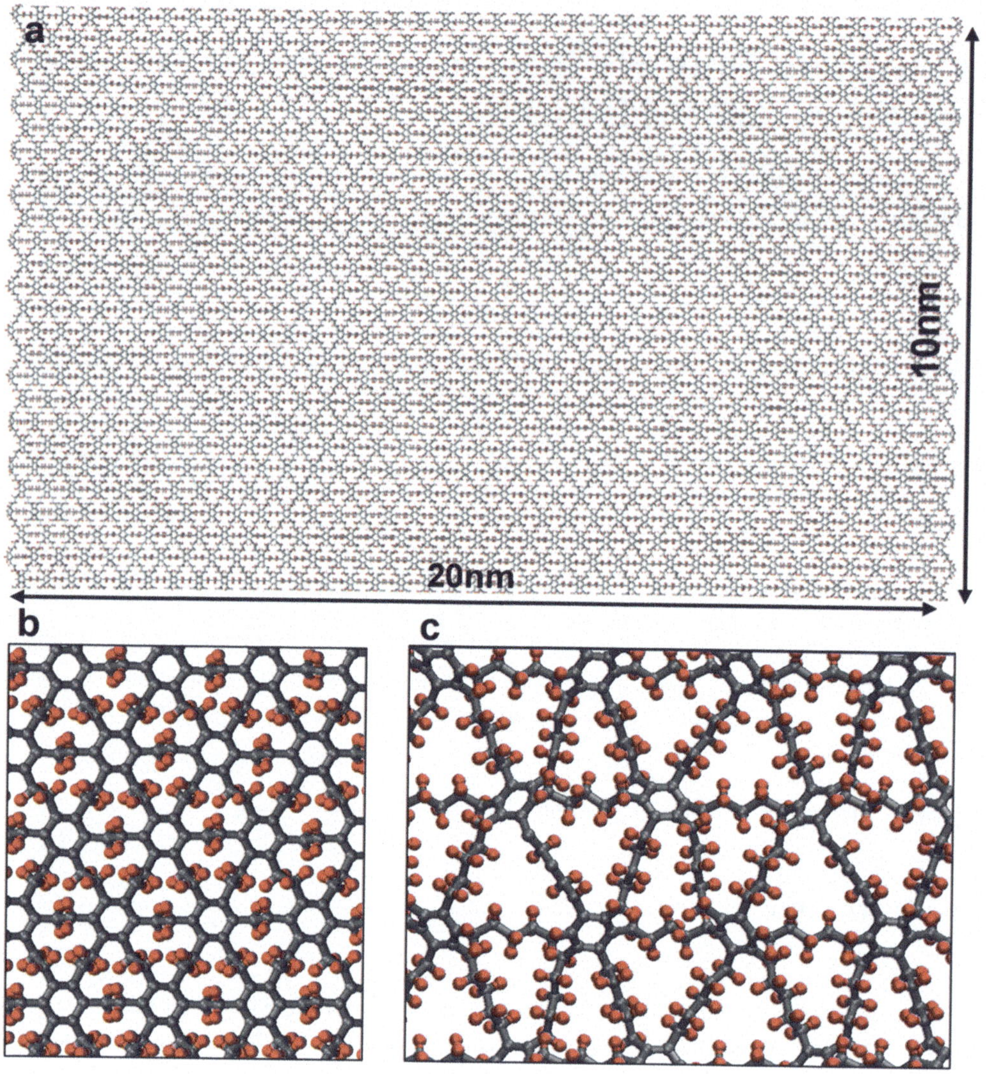

Fig. 5.1 (**a**) Domain for mechanical MD simulations with strain applied in the vertical direction, (**b**) GrE-2 structure (grey atoms depict carbon, red atoms depict hydrogen for ease of viewing) and (**c**) GrE-5 structure with same color scheme showing out-of-plane tilt in the benzene rings

The engineering stress-strain plots for GrE-2 through GrE-6 can be seen in Fig. 5.2a, and the 2D elastic modulus and strength are summarized in Fig. 5.2b, c respectively. The elastic modulus decreases non-monotonically from GrE-2 to GrE-6, with even numbered GrE-n structures having higher moduli than their even numbered GrE-n counterparts. This discrepancy is not a result of a fundamental change in the direction of strain, since all structures were strained along the same crystalline direction. While the inverse correlation between elastic modulus and bridge chain length is clear, there is another underlying cause for the discrepancy between odd and even numbered GrE-n structures. All of the simulated odd GrE-n structures show benzene rings with significant tilt (20–30°) out of the principal plane of the 2D polymer in their lowest-energy configuration, caused by the arrangement of the polyethylene bridge units. Even numbered GrE-n structures have bridge units allow for a configuration where each of the six bridge units on a given node alternates their out-of-plane zig zag orientation around the same benzene ring, preserving symmetry (Fig. 5.1b). The odd numbered GrE-n structures cannot preserve the same symmetry, and thus cause the benzene rings to tilt out of plane to accommodate the bridge units (Fig. 5.1c). The tilt to the benzene rings significantly increases the compliance of the structures, leading to far lower elastic moduli. While the stiffness of GrE-n is strongly impacted by the tilt in the benzene rings, the strength is less sensitive to the out-of-plane tilt between odd and even numbered bridge units (Fig. 5.2c). The added compliance does not weaken the structures, leading to the odd-numbered structures breaking at relatively high strains.

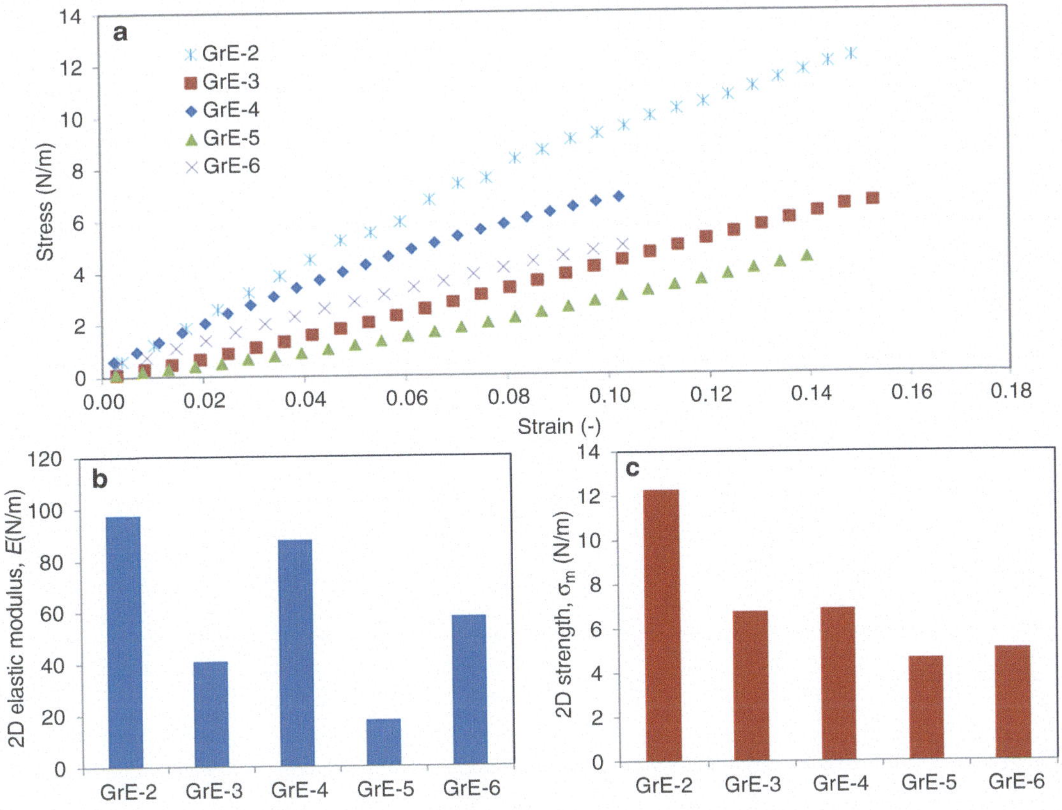

Fig. 5.2 (**a**) 2D stress-strain plot, (**b**) elastic modulus and (**c**) strength for varying graphylene structures

Designing 2D polymers to have added compliance through nodal tilt could allow for relatively facile tailoring of the mechanical properties to suit application needs, such as having a given 2D polymer for a high strength, low strain application, and an entirely different 2D polymer for a high strain-energy application. The added compliance can also be used to increase the fracture toughness of these materials, opening a wide array of possibilities for customizing the mechanical properties.

Chapter 6
Mechanical Behavior of Additively Manufactured Ti-6Al-4V Following a New Heat Treatment

Jonathan P. Ligda, Brady G Butler, Nathaniel Saenz, and James Paramore

Abstract Hydrogen sintering and phase transformation is a Ti-6Al-4V heat treatment process capable of normalizing the microstructure of bulk parts. The mechanical properties of these processed parts are comparable to that of wrought Ti-6Al-4V, which makes it attractive for use in manufacturing areas where the resulting microstructure is difficult to control, like powder metallurgy and additive manufacturing. To investigate the application space of this heat treatment, Ti-6Al-4V parts were printed from three different additive methods (DMLS, EBM, and cold spray) and their tensile properties were evaluated in both the as-printed and heat treated states. Due to the size of the printed parts, millimeter scale tensile specimens were used and care must be taken to ensure the reduced sample size still produces reliable results. Preliminary tension tests show that the heat treatment process normalizes the microstructure, closes porosity, and improves ductility.

Keywords Additive manufacturing · Ti-6Al-4V · Mechanical properties · Small-scale

The near-net shape fabrication and access to complex geometries makes additive manufacturing an attractive technique for creating metal parts [1]. There continues to be improvements for decreasing porosity and controlling microstructure. However, printing Ti-6Al-4V parts is difficult due to the need to limit oxygen content, and is usually overcome by printing in a controlled argon atmosphere [2]. Following printing, a hot isostatic pressing (HIP) step can be done to decrease some of the porosity and improve the mechanical properties [3, 4]. Hydrogen sintering and phase transformation (HSPT), a process originally developed for sintering Ti-6Al-4V powders, can also be used to improve the mechanical properties for bulk parts [5]. The HSPT process reduces porosity similar to the HIP, but is also capable of controlling the final microstructure of the part [6]. This means that strength and ductility of a printed any part can be improved regardless of the starting microstructure.

Ti-6Al-4V parts were printed using three different types of additive manufacturing methods, direct metal laser sintering (DMLS), electron beam melting (EBM), and cold spray (CS). Each method has a different resulting microstructure and shows drastically different mechanical properties. Parts from each method were processed via HSPT at temperatures of 1050–1200 °C for 1–8 h depending on the starting porosity. Millimeter scale tension specimens with gauge sections of $0.7 \times 0.5 \times 5$ mm were machined along the build and raster directions of each method. Tension tests were performed using a custom built testing rig which had a 250 lb load cell, DC stepper motor, and 12 megapixel camera. Images captured during the tension test were used for digital image correlation (DIC) based strain calculations.

Figure 6.1a shows the engineer stress vs. strain curves for the DMLS and DMLS+HSPT samples taken from the raster direction. Their strengths are comparable, with a slight increase for HSPT processed sample. There is also an increase in the total elongation of the HSPT sample, 6%, higher than the DLMS only elongation of 3%. Even those this is higher, it is still much lower than the expected ductility found in wrought Ti-6Al-4V. Scanning electron microscope (SEM) images inset in the plot show the fracture surfaces of these two samples. There is no noticeable difference between the two. Figure 6.1b shows the engineering stress vs. strain curves for the CS, CS+hot isostatic pressed (HIP), and CS+HSPT samples. The CS only sample consists mostly of cold welded spheres and still contains a large amount of porosity, as shown by the fracture surface shown in the inset SEM image. Under tension, the sample fractures at a stress level under 200 MPa, significantly lower than that of

J. P. Ligda (✉) · B. G. Butler
US Army Research Laboratory, Adelphi, MD, USA
e-mail: jonathan.p.ligda.civ@mail.mil

N. Saenz
Army Educational Outreach Program, Concord, NH, USA

J. Paramore
Oak Ridge Associated Universities, Oak Ridge, TN, USA

S. Kramer et al. (eds.), *Mechanics of Additive and Advanced Manufacturing, Volume 8*, Conference Proceedings of the Society for Experimental Mechanics Series, https://doi.org/10.1007/978-3-319-95083-9_6

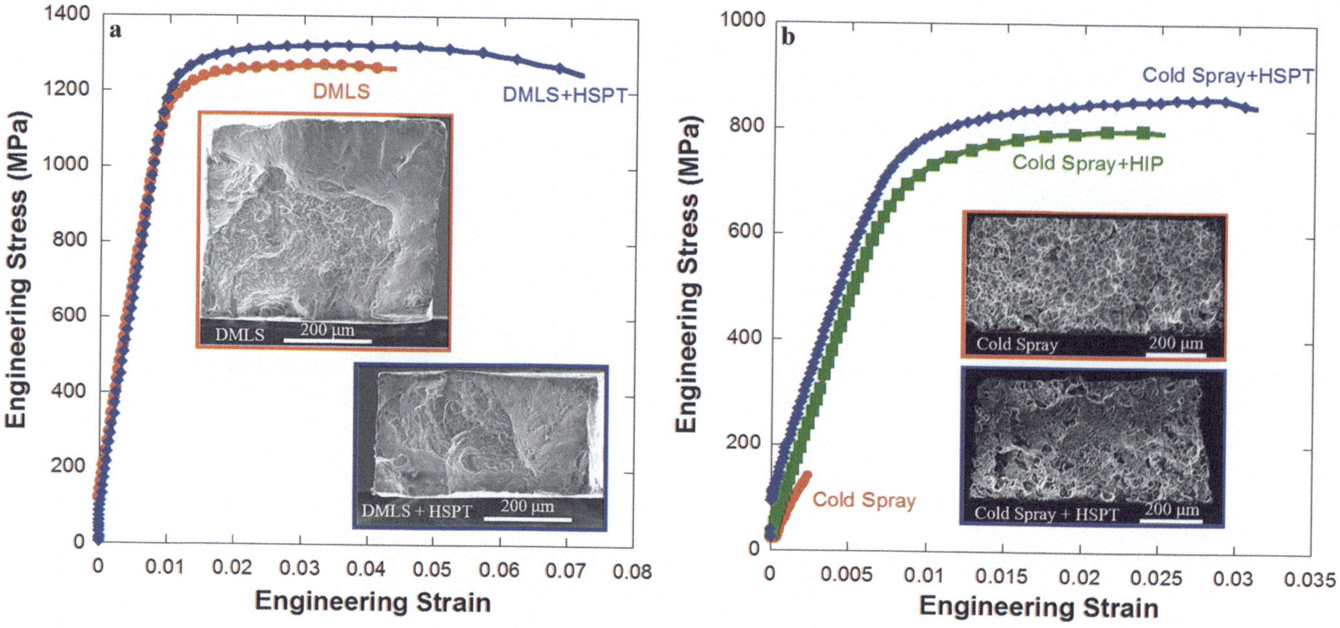

Fig. 6.1 (**a**) Engineering stress vs. strain plot of the DMLS and DMLS+HSPT with inset scanning electron microscopy images of the fracture surface. (**b**) Engineering stress vs. strain plot of CS, CS+HIP, and CS+HSPT with inset scanning electron microscopy images of the fracture surface

wrought Ti-6Al-4V. For the CS+HIP sample, the HIP process greatly improves the strength and partially improves the elongation to near 1.6%. The CS+HSPT sample shows a slightly improved strength and increased elongation to 2.2%. Inset SEM image of the fracture surface show the improved densification of the samples following HSPT, but large pores still remain. This contributes significantly to the low ductility.

References

1. Frazier, W.E.: Metal additive manufacturing: a review. J. Mater. Eng. Perform. **23**(6), 1917–1928 (2014)
2. Xu, W., Brandt, M., Sun, S., Elambasseril, J., Liu, Q., Latham, K., Xia, K., Qian, M.: Additive manufacturing of strong and ductile Ti–6Al–4V by selective laser melting via in situ martensite decomposition. Acta Mater. **85**, 74–84 (2015)
3. Leuders, S., Thöne, M., Riemer, A., Niendorf, T., Tröster, T., Richard, H.A., Maier, H.J.: On the mechanical behaviour of titanium alloy TiAl6V4 manufactured by selective laser melting: fatigue resistance and crack growth performance. Int. J. Fatigue. **48**, 300–307 (2013)
4. Das, S., Wohlert, M., Beaman, J.J., Bourell, D.L.: Processing of titanium net shapes by SLS/HIP. Mater. Des. **20**(2), 115–121 (1999)
5. Sun, P., Fang, Z.Z., Koopman, M.: A comparison of Hydrogen Sintering and Phase Transformation (HSPT) processing with vacuum sintering of CP-Ti. Adv. Eng. Mater. **15**, 10 (2013)
6. Paramore, J.D., Fang, Z.Z., Sun, P., Koopman, M., Chandran, K.S.R., Dunstan, M.: A powder metallurgy method for manufacturing Ti-6Al-4V with wrought-like microstructures and mechanical properties via hydrogen sintering and phase transformation (HSPT). Scr. Mater. **107**, 103–106 (2015)

Chapter 7
Dynamic Thermal Softening Behavior of Additive Materials for Hybrid Manufacturing

Steven Mates, Mark Stoudt, Gregor Jacob, Wilfredo Moscoso, and Vis Madhavan

Abstract Hybrid manufacturing involves both additive and subtractive (machining) processes to achieve the final product. Substantial differences can exist between the mechanical behavior of additively as-built materials compared to their wrought counterparts. As such, the use of wrought material properties for the simulation and optimization of the machining step in a hybrid manufacturing process may produce inaccurate results. The present work uses the NIST pulse-heated compression Kolsky bar to measure the dynamic behavior of both wrought and additively produced Inconel 625 and 17-4 PH stainless steel over a range of temperatures up to 1000 °C and at strain rates of 3000 s^{-1}. The measurement results are correlated to underlying microstructural differences between additive and wrought materials that arise because of the differences between these material processing routes as described in the literature.

Keywords Kolsky bar · Additive manufacturing · Hybrid manufacturing · High strain rate · High heating rate

7.1 Introduction

Hybrid manufacturing combines additive manufacturing (AM) with traditional machining to improve the surface quality of additively made parts while preserving the benefits of the AM route. AM is particularly attractive for advanced, high-value materials and parts that are costly or more difficult to produce using traditional methods. Inconel 625 (IN 625) and 17-4 PH stainless steel are two alloys of active interest in the AM research community. The microstructure of as-built AM materials can be very different from their wrought (cast then thermo-mechanically processed) counterparts, primarily due to differences in the thermal history experienced by the material during processing. In 17-4 PH stainless steel, the AM material can contain large amounts of retained austenite [1], whereas the wrought form is entirely martensitic. AM IN 625 contains brittle precipitates [2] that are intentionally avoided in wrought products by careful control of thermal history during traditional processing. These significant microstructural differences likely alter the mechanical behavior of, and, potentially, the machining performance of, AM materials compared to wrought. The machining behavior of AM alloys is currently under investigation at Wichita State University, and the results of this research will be presented in a companion paper at this conference. In this work they found notable differences in cutting temperature, chip morphology and tool wear between wrought and AM IN 625.

The present work compares the dynamic (strain rate $\approx$ 3000 s^{-1}) plastic response of additively-manufactured IN 625 and 17-4 PH stainless steel against their wrought equivalents at temperatures up to 1000 °C using the NIST pulse-heated Kolsky bar [3]. Short (sub 4 s) total heating times are employed to approach thermal conditions during machining, where heating rates are quite rapid such that equilibrium microstructures may not have time to evolve despite the elevated temperature. In 17-4 PH stainless steel, for example, mechanical properties can change after as little as 30 s of exposure to elevated temperatures [4], presumably due to coarsening of fine Cu precipitates [5].

S. Mates (✉) · M. Stoudt · G. Jacob
National Institute of Standards and Technology, Gaithersburg, MD, USA
e-mail: steven.mates@nist.gov

W. Moscoso · V. Madhavan
Wichita State University, Wichita, KS, USA

S. Kramer et al. (eds.), *Mechanics of Additive and Advanced Manufacturing, Volume 8*, Conference Proceedings of the Society for Experimental Mechanics Series, https://doi.org/10.1007/978-3-319-95083-9_7

7.2 Experimental

Additively manufactured IN 625 and 17-4 PH stainless steel were produced by a commercial powder bed fusion device. In this process, a laser is rastered over a powder layer, selectively melting material in the desired pattern to solidify onto the previously built layer. The process is then repeated with a freshly laid powder layer. Once completed, the loose powder is removed and the builds are subject to manufacturer-recommended heat treatment schedules which vary depending on the alloy. Compression specimens measuring 4 mm in diameter by 2 mm thick were then cut from the cooled and stress-relieved build materials via electron discharge machining (EDM). The specimens were cut such that the compression direction is perpendicular to powder layer and parallel to the build direction. Samples of the companion wrought material were cut from square stock, also using EDM. Both additive and wrought IN 625 specimens were subject to an additional annealing of 1038 °C (1900 °F) for 5 min followed by an air cool.

Samples were dynamically compressed using the NIST pulse-heated Kolsky bar [3]. For elevated temperature testing, samples are subjected to a single, short direct current pulse at low voltage provided by a large battery bank (13.2 V). Heating rates of 1000 °C/s or more are possible with this technique, as is precise control of the shape and duration of the thermal history, using feedback-control on either the current or the temperature, the latter as measured via infrared pyrometry. Graphite foils were used between the sample and the bars to eliminate any arcing at the sample-bar interface during heating. Stress-strain data were obtained from the incident and transmitted strain wave pulses, measured at 2 MHz, in the usual manner [6], with two noteworthy exceptions. First, the foil mechanical behavior was modeled and subtracted from the overall response to extract the specimen stress-strain behavior for pulse-heated experiments. Details of the foil correction method are given in reference [3]. Second, we employed an elastic punching correction [7] to account for elastic indentation of the bar faces for both heated and room temperature tests. Finally, mild pulse shaping was used to smooth out oscillations associated with the striker impact.

7.3 Results

Figure 7.1 shows room temperature high strain rate flow stress curves of additive and wrought IN 625 and 17-4 PH stainless steels. Note that the downturns at the ends of the stress-strain curves in Fig. 7.1 are caused by the release-wave that ends the dynamic compression event, and do not represent specimen failure. In both plots, nominal linear elastic responses are shown for each material in the wrought condition for comparison. The good agreement between the initial portion of the dynamic stress-strain curves measured here with the corresponding nominal elastic responses was substantially aided by using the elastic punch correction technique [7]. AM IN 625 exhibits higher flow stress but similar strain hardening up to about 25%

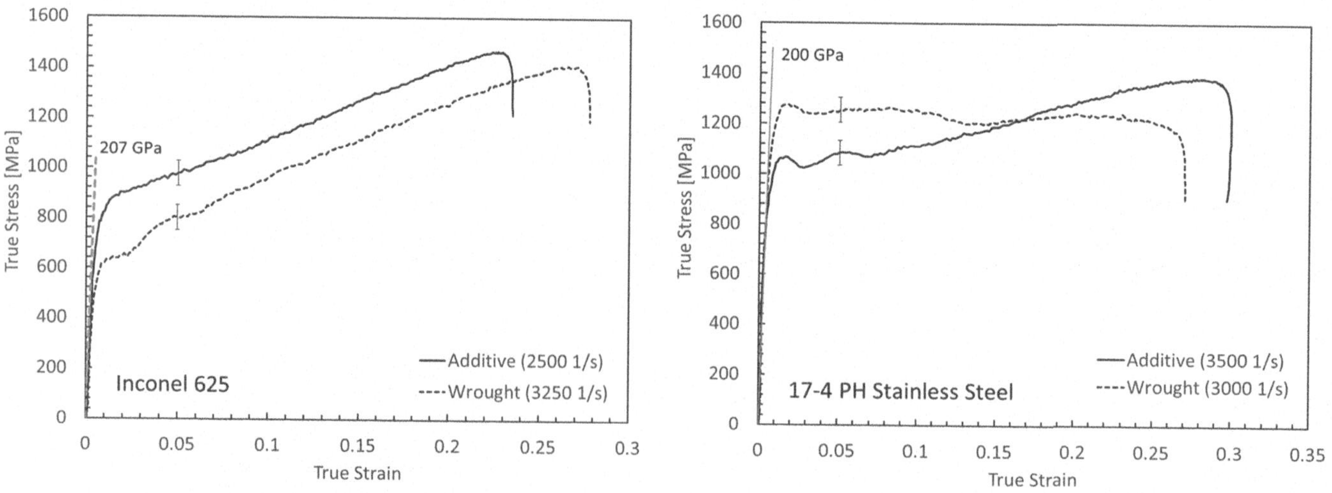

Fig. 7.1 Comparison of room temperature high strain rate compressive response of additive and wrought IN 625 (left) and 17-4 PH stainless steel (right). Linear elastic responses are drawn for comparison using nominal wrought Young's modulus values from commercial literature. The average strain rate for the tests are noted parenthetically in the legends. Error bars are computed using random error propagation and apply uniformly to the entire curve

strain. Conversely, AM 17-4 PH stainless steel shows a lower apparent dynamic yield stress but a significantly higher strain hardening rate compared to the wrought material. Both observations can be readily explained by differences in the microstructures between additively-produced and wrought materials. Wrought 17-4 SS is mostly martensite, resulting in a very high yield strength with little strain hardening capacity. Additively manufactured 17-4 steel, by contrast, can contain significant retained austenite. The retained austenite is believed to occur due to a combination of enriched austenite-stabilizing elements, including excess nitrogen absorbed during deposition or during atomization of feedstock powder, and a finer solidification microstructure resulting from the high cooling rates associated with AM layer solidification [1]. Retained austenite, being weaker than martensite, reduces the hardness and yield strength of the bulk AM material compared to the wrought material. In addition, the presence of FCC allows much greater strain hardening capacity than its wrought (BCC martensite) counterpart. For IN 625, both AM and wrought forms consist of single-phase FCC Ni, yet AM IN 625 contains a significant amount of Ni_3Nb, a brittle intermetallic phase, referred to as δ-phase, which forms during stress-relief annealing. This could explain the observed increased hardness and lower ductility compared to wrought [2]. Additional strengthening mechanisms include grain size refinement in the AM material due to the high inherent cooling rates [2] and by increased initial dislocation density [8]. As Fig. 7.1 shows, the dynamic flow stress of the AM IN 625 is higher compared to wrought at room temperature. However, the strain-hardening behavior of the two materials is quite similar, which is consistent with the fact that both are predominantly FCC Ni.

Figure 7.2 shows the plastic flow stress of AM and wrought forms of both materials at several temperatures up to about 1000 °C at a fixed level of plastic strain. In the IN 625 system, the AM material remains stronger than the wrought material at almost all temperatures, and the thermal softening behavior generally mirrors that of the wrought material. This is somewhat expected, given that the brittle δ-phase likely present in the AM material is stable to well over 1000 °C, and is relatively difficult to dissolve [2], requiring heating times much longer than the 3.5 s used here. The δ-phase provides an additional athermal barrier to slip and remains over the entire temperature range examined, boosting the flow stress at all temperatures. Suppression of grain coarsening or recrystallization on heating by the limited heating times also likely contributes to the persistence of the strength offset between AM and wrought observed even at high temperatures. The flow stress of 17-4 PH stainless steel is plotted against temperature in this figure at a smaller amount of plastic strain (5%) to highlight the initial reduction of the yield strength due to significant retained austenite in the AM material. Because the AM material strain hardens much more readily than the wrought at room temperature, comparisons of thermal softening behavior made at a single strain level will change based on strain level chosen. At 5% strain, the wrought material maintains its strength advantage over the AM material until near the temperature at which martensite starts to transform to austenite, which for this alloy is about 750 °C [1]. Above this temperature, the flow stresses of AM and wrought appear to fall in line with each other, which is expected given that both materials should be fully austenitic given sufficient heating time. More data in this region are needed to confirm this and to explore how the kinetics of this phase transformation affect the flow stresses of these materials near this transformation temperature.

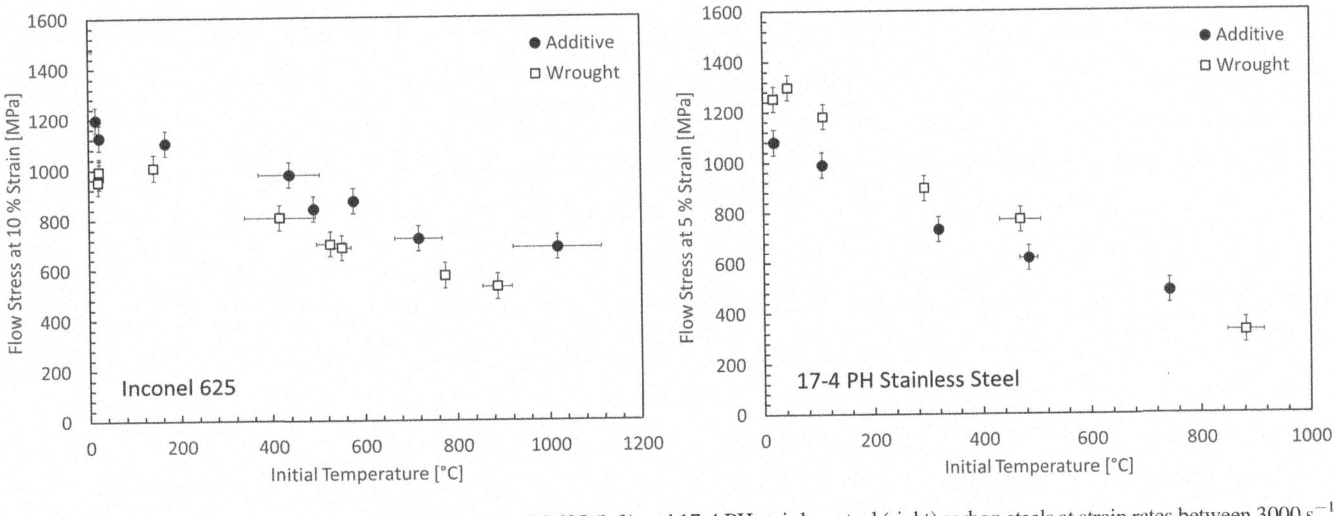

Fig. 7.2 Thermal softening behavior of AM and wrought IN 625 (left) and 17-4 PH stainless steel (right) carbon steels at strain rates between 3000 s^{-1} and 5000 s^{-1}. Error bars on stress are computed from error propagation (95% confidence interval) and on temperature are computed from thermal gradients measured via pyrometer readings obtained on opposite sides of the specimen

Differences in the dynamic plastic response between AM and wrought materials may translate into differences in machining behavior of AM builds compared to their wrought equivalents. Comparative cutting tests between AM and wrought IN 625 have shown that the AM form produces higher chip temperatures and a more irregular chip segmentation compared to the wrought. This agrees with the trends in the current data that show increased flow stress at all temperatures and, one assumes, lower ductility due to the presumed presence of δ-phase. Comparative cutting tests with additive and wrought 17-4 PH stainless steel are planned, and these tests are expected to reveal the effects of lower yield and larger strain hardening capacity of the AM material.

7.4 Conclusions

Hybrid manufacturing, which involves secondary machining operations on additively manufactured materials, hinges on the ability to effectively machine these new materials, which tend to have very different microstructures in the as-built condition compared to typical wrought materials of the same composition. The high strain rate plastic behavior of AM IN 625 and 17-4 PH stainless steel is shown to be significantly different at elevated temperatures, and in the case of the IN 625, the results are congruent with comparative machining tests performed at Wichita State University which show measurable differences in cutting temperature, chip morphology and tool wear between wrought and AM IN 625.

References

1. Cheruvathur, S., Lass, E.A., Campbell, C.E.: Additive manufacturing of 17-4 PH stainless steel: post-processing heat treatment to achieve uniform reproducible microstructure. JOM. **68**(3), 930–942 (2016)
2. Lass, E.A., Stoudt, M.R., Williams, M.E., Katz, M.B., Levine, L.E., Phan, T.Q., Gnaeupel-Herold, T.H., Ng, D.S.: Formation of the Ni_3Nb δ-phase in stress-relieved Inconel 625 produced via laser powder-bed fusion additive manufacturing. Metall. Mater. Trans. A. **48**(11), 5547–5558 (2017)
3. Mates, S.P., Rhorer, R., Whitenton, E., Burns, T., Basak, D.: A pulse-heated Kolsky bar technique for measuring the flow stress of metals at high loading and heating rates. Exp. Mech. **48**, 799–807 (2008)
4. Steel, A.K.: 17-4 PH Stainless Steel: Product Data Bulletin, AK Steel Corporation, West Chester, OH, USA (2015)
5. Murayama, M., Katayama, Y., Hono, K.: Microstructural evolution in a 17-4 PH stainless steel after aging at 400°C. Metall. Mater. Trans. A. **30A**, 345–353 (1999)
6. Song, B., Chen, W.W.: Split Hopkinson (Kolsky) Bar: Design, Testing and Applications. Springer, Boston (2011)
7. Safa, K., Gary, G.: Displacement correction for punching at a dynamically loaded bar end. Int. J. Impact Eng. **37**, 371–384 (2010)
8. Lyle Levine, NIST, private communication

Chapter 8
Correlation Between Process Parameters and Mechanical Properties in Parts Printed by the Fused Deposition Modeling Process

Samuel Attoye, Ehsan Malekipour, and Hazim El-Mounayri

Abstract Fused deposition modeling (FDM) represents one of the most common techniques for rapid prototyping and industrial additive manufacturing (AM). Optimizing the process parameters which significantly impact the mechanical properties is critical to achieving the ultimate final part quality sought by industry today. This work investigates the effect of different process parameters including nozzle temperature, printing speed, and print orientation on Young's modulus, yield strength, and ultimate strength of the final part for two types of filament, namely, Poly Lactic Acid (PLA) and Acrylonitrile Butadiene Styrene (ABS). Design of Experiments (DOE) is used to determine optimized values of the process parameters for each type of filaments; also, a comparison is made between the mechanical properties of the parts fabricated with the two materials. The results show that Y-axis orientation presents the best mechanical properties in PLA while X-axis orientation is the best orientation to print parts with ABS.

Keywords Additive manufacturing · Fused deposition modeling · Process parameters · Mechanical properties

8.1 Introduction

Fused deposition modeling (FDM) is a subset technology of additive manufacturing by material extrusion. FDM is devised and patented by Scott Crump [1] and Industrial Technology Research Institute (ITRI)[1] stated that this emerging technology, represents one of the most common techniques for prototyping and personal additive manufacturing (AM). FDM also produces end-use parts fit for mechanical, chemical and biochemical operating conditions. This technology, however, is not free from limitation. In this process there is an instant hardening of the plastic as it bonds to the layer below and another layer is deposited by the extrusion nozzle. In this procedure, the properties of the deposited material, the printer specifications especially in 3D printing finely-detailed items, and the range of process parameters affect significantly on the finished product quality. Any depletion in these categories leads to some potential defects such as unfavorable mechanical properties, poor surface condition, etc. Thus, scholars have focused in these fields to improve the ultimate quality of the fabricated parts. In an effort, Dorigato et al. [2] included reinforcing particles (i.e., Nano-fillers) with the polymer to improve the mechanical properties of FDM final products. It was observed that this approach significantly boost the properties such as impact resistance, elastic modulus, thermal stability, and fire resistance, compared with those of neat polymers [2]. In a different effort, Keng Hsu at the Arizona State University improved the interfacial strength by pre-heating the base layer [3]. On the other hand, optimization of the process parameters is considered an important alternative for improving the quality of the final parts. Several efforts are underway to improve the mechanical properties of FDM parts; for an instance, in a study by Onwubolu and Rayegani, a functional relationship between process parameters and tensile strength for FDM process was developed using group method for data modeling. In this study, the effect of layer thickness, part orientation, raster angle, raster width, and air gap on the tensile strength of test specimen was studied [4]. In another study, Stephen A., et al. (2014) identified the tests that can be most sensitive to the changes in processing conditions and differences in interlayer bond strength which affect the quality of fabricated parts. The various mechanical properties of tensile, flexural properties, notched and un-notched impact strengths and the influence of changes in processing parameters and orientations of specimens on the quality was investigated. Analysis of variance (ANOVA) was used to compare the significance of the effect of processing

[1]https://www.itri.org.tw

S. Attoye · E. Malekipour (✉) · H. El-Mounayri
Collaborative Additive Manufacturing Research Initiative at IUPUI (CAMRI), Purdue School of Engineering and Technology, Indianapolis, IN, USA
e-mail: soattoye@umail.iu.edu; emalekip@purdue.edu; helmouna@iupui.edu

S. Kramer et al. (eds.), *Mechanics of Additive and Advanced Manufacturing, Volume 8*, Conference Proceedings of the Society for Experimental Mechanics Series, https://doi.org/10.1007/978-3-319-95083-9_8

parameters on the mechanical properties. Their conclusion, on the basis of the data from experiments, was that a combination of measures of modulus, weight and impact strength can be used to infer the quality of a build on an FDM machine [5]. Previous works investigated the effects of different materials or changing process parameters on mechanical properties; however, there are still gaps in literature relating to the comparison between the obtained mechanical properties for the part printed by prevalent materials in FDM process and also, introducing the optimized values for process parameters which lead to the best mechanical properties for the parts printed by each material. In this work, we investigate the influence of printing orientation, nozzle temperature, and printing speed on Young's modulus, yield strength, and ultimate strength of a modified ASTM tensile strength test standard part (ASTM D638) [6] for two types of material, namely, Poly Lactic Acid (PLA) and Acrylonitrile Butadiene Styrene (ABS), used predominantly in FDM process. Printing orientation plays a significant role in the ultimate mechanical properties since it affects the bonding between layers. As a matter of fact, the anisotropy of the fabricated parts has been always a fundamental problem in FDM and thus, selecting the right printing orientation according to the function of the part is crucial. The nozzle temperature also affects the quality of the part directly. This temperature was adjusted based on the recommended temperature on each type of filament rolls. Effects of changing in printing speed are also investigated and depicted in our work. As it already mentioned, we conducted our experiments for both PLA and ABS.

8.2 Methodology and Experimental Setup

The interaction between printing parameters (nozzle temperature, printing speed, and print orientation) adjusted during fabrication and the mechanical properties (Young's modulus, yield strength, and ultimate tensile strength) was observed. The specimen selected and modified (Fig. 8.1) for monitoring and evaluation is based on the ASTM tensile strength test standard part (ASTM D638). The dimensions followed by minor adjustments to facilitate the printing process.

Two sets of the specimen were fabricated. Set-A consisted of thirty-six (36) individual printed MakerBot PLA specimens (specimen A1 to A36). Set-B consisted of twenty-seven (27) individual printed Filabot ABS specimens. The printing process parameters (nozzle temperature, printing speed), adjusted for each part, are listed in Table 8.1. The printing parameter values were selected based on the machine and material specifications. Three nozzle temperature values (low, medium and high) were considered for each filament according to the recommended temperature of the filament rolls. Various build orientations used in the experiments are shown in Fig. 8.2. The X, Y, Z orientations were used in the fabrication of both set-A and set-B specimens; while the 45-degree orientation was only used for set-A specimens (Fig. 8.2).

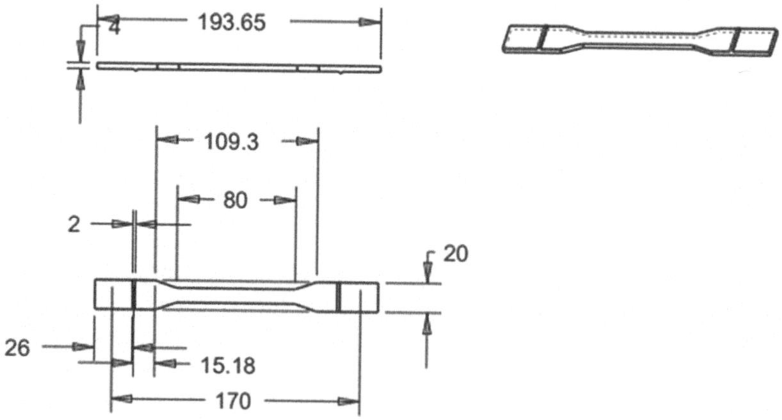

Fig. 8.1 The modified ASTM tensile strength sample used for experiments

Table 8.1 Printing parameters adjusted for fabrication process of set-A specimen A1-A36 and set-B specimen B1-B27

Printing parameters	Values for printer A			Values for printer B		
Nozzle temperature in °C	200	215	220	215	225	235
Printing speed in mm/s	20	40	60	20	40	60

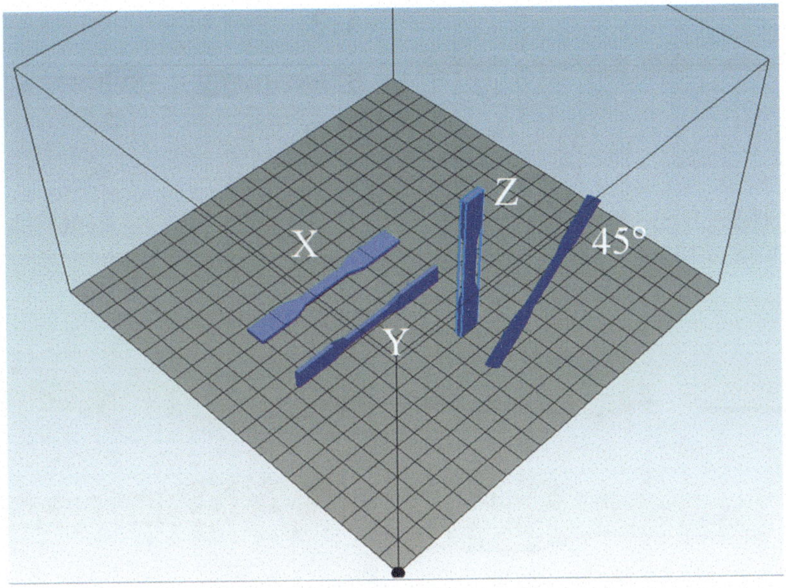

Fig. 8.2 X, Y, Z, and 45° orientation used in fabrication of specimen set-A and X, Y, and Z orientation used in fabrication of set-B

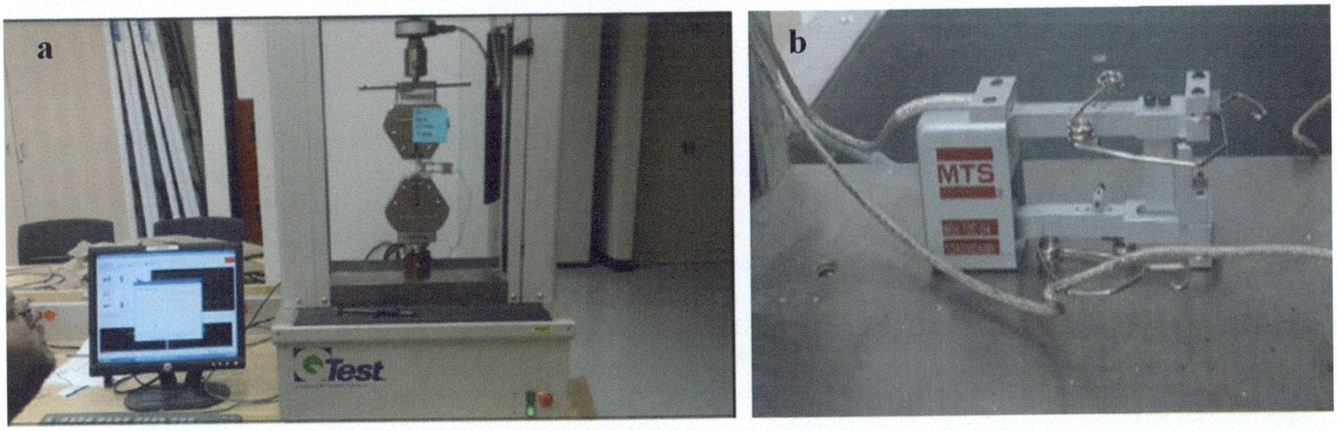

Fig. 8.3 (**a**) Q-Test machine (**b**) Matt Struve Demo MTS Extensometer

After the specimens were printed, the tensile strength test was performed on each part. The results were analyzed and revealed in Sect. 8.3. Mechanical analysis of set-A and set-B was performed using Q-test machine, Matt Struve Demo MTS Extensometer, and test-works software. Young's modulus, yield strength, and ultimate strength of the FDM printed specimens were obtained.

8.3 Results and Discussion

During the printing process, it was observed that the specimens fabricated in the X orientation took the shortest time to print for both set-A and set-B specimens. This was due to the fewer layers required to complete the part built in this orientation. For set-A specimens, printing at the 45-degree orientation took the longest time due to the requirement of support structures. On the other hand, printing at the Z-axis orientation took the longest time for set-B specimens. This was due to the requirement of support structures and the specimen height. The effect of the printing speed on the build time was not significant; however, the mechanical analysis shows a more discernable influence of the printing speed parameter on the build process and final product mechanical properties.

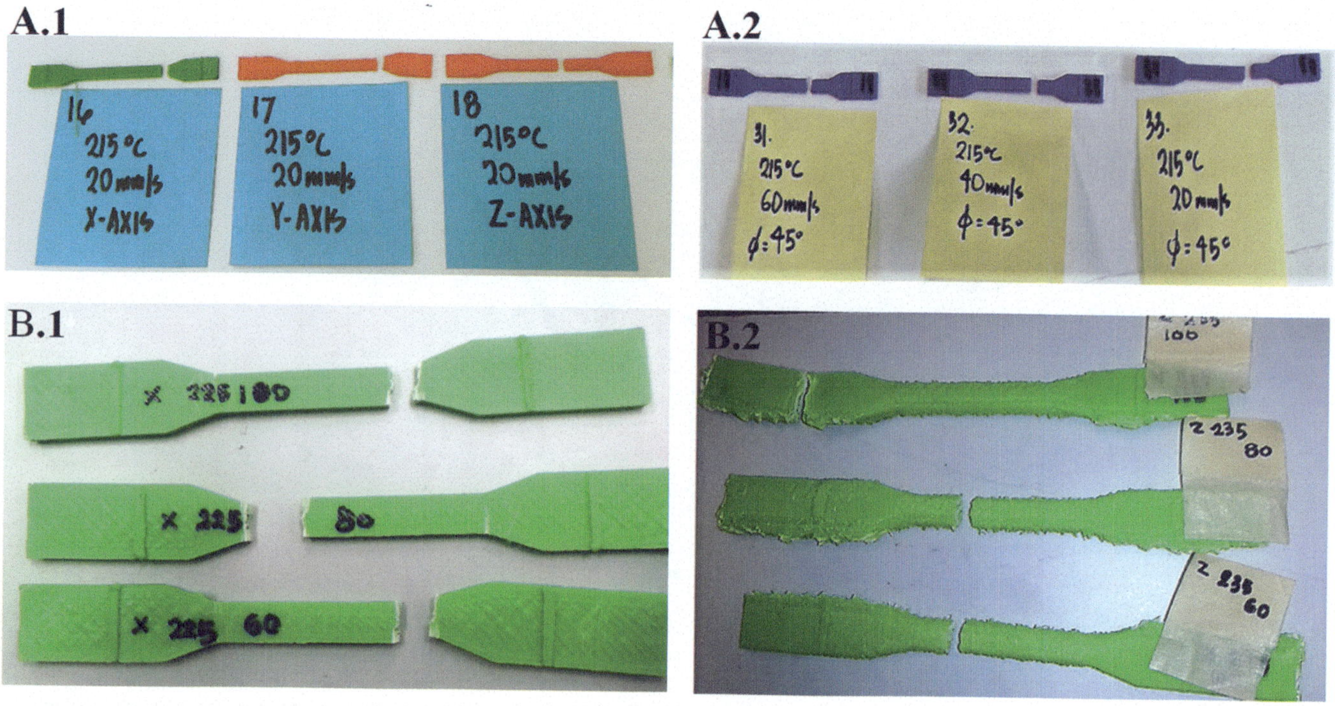

Fig. 8.4 A.1 specimens printed in X, Y, Z orientation with PLA; A.2 specimens printed in 45° with PLA; B.1 specimens printed in X orientation with ABS; B.2 specimens printed in Z orientation with ABS

Almost all the specimens printed with either PLA or ABS (set-A or set-B) in the X and Y orientations tended to fracture at location close to the end of the sample, at the "neck" region; however, around 33% of the specimens in both sets printed in the z-orientation, tended to fracture at location that is closer to the mid-region of the specimens and 66% close to the end of the samples. In the specimens printed at 45-degree in set-A, the fracture spot tends to shift to the middle for all the process parameters. Some specimens after evaluation on the Q-test machine are depicted in Fig. 8.4.

The effects of changing the printing parameters, i.e. nozzle temperature, printing speed, and printing orientations, on the mechanical behavior of each specimen are investigated. In this regard, values of Young's modulus, yield strength, and ultimate tensile strength of the printed specimens in X, Y, Z, and 45-degree directions for PLA filament and X, Y, and Z for ABS filament for different nozzle temperatures and printing speeds gotten from the tensile strength and are presented in the Sects. 8.3.1 and 8.3.2.

8.3.1 Mechanical Properties of the Specimens Fabricated with PLA Filament in X, Y, Z, and 45°

Figures 8.5, 8.6, and 8.7 show respectively Young's modulus, yield strength, and ultimate tensile strength of the fabricated PLA (set-A) specimens. It can be observed from Figs. 8.5, 8.6, and 8.7 that, the specimens printed in the Y–axis orientation show the highest mechanical properties compared with the specimens printed in the X, Z -axis and 45-degree orientation. In these samples, the minimum value of mechanical properties in Y-direction is predominantly higher than the maximum one in the other directions. Therefore, Y is the best direction suggested for printing with PLA. For all the mechanical properties, 45-degree direction shows the weakest properties; thus, this direction is not recommended for printing of parts. The diagrams also depict that the highest values of almost all the mechanical properties were obtained at the lowest temperature which is 200 °C and the highest printing speed which is 60 mm/s for all X, Y, and Z orientations. Increasing the temperature weakens the mechanical properties for this speed. Changing the nozzle temperature does not show any specific trend on changing of mechanical properties for the other printing speeds.

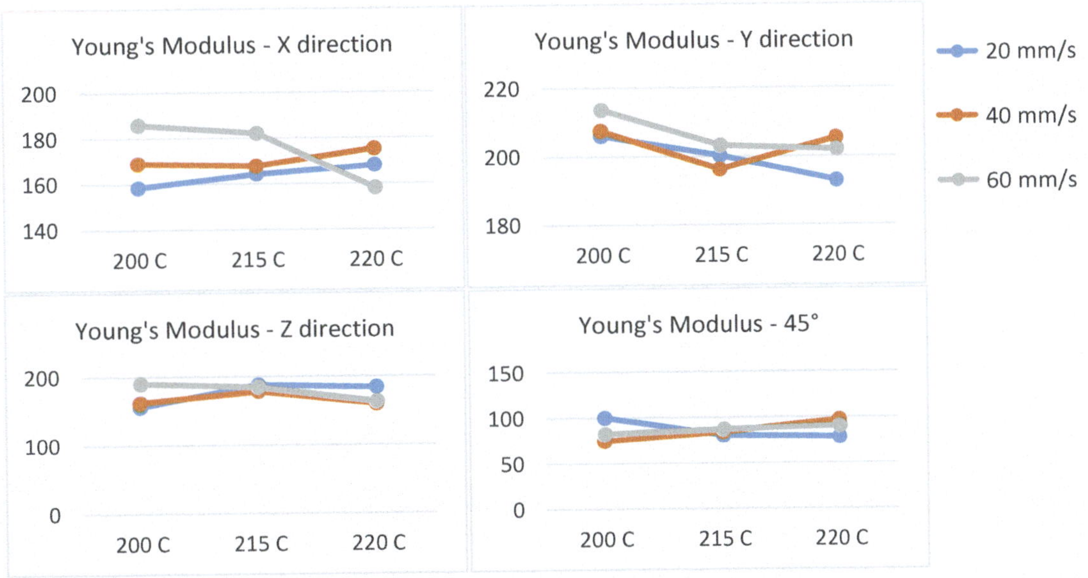

Fig. 8.5 Young's modulus in X, Y, Z orientations and 45° for the part printed with PLA in different nozzle temperatures and printing speeds

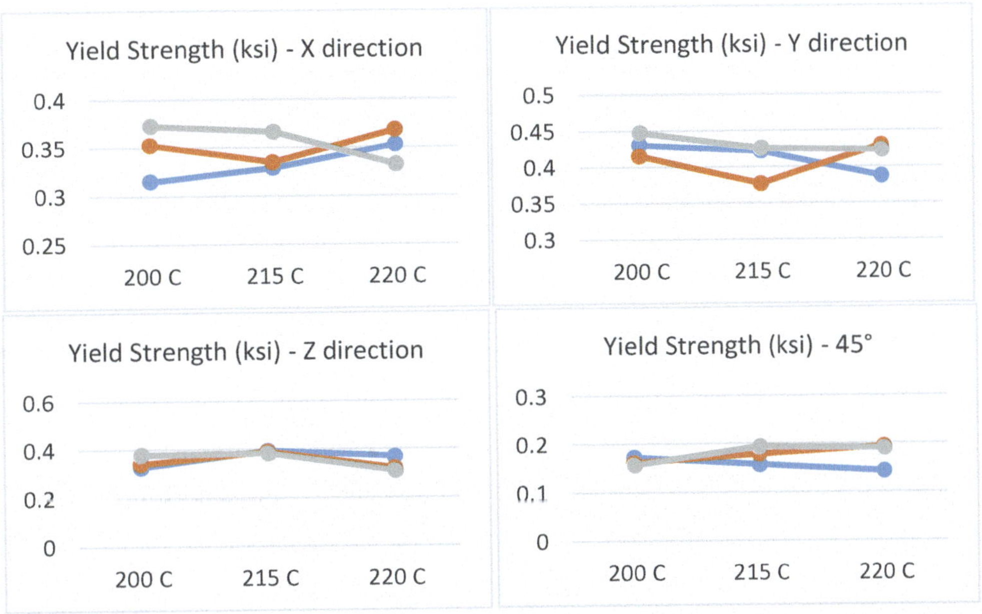

Fig. 8.6 Yield strength in X, Y, Z orientations and 45° for the part printed with PLA in different nozzle temperatures and printing speeds

8.3.2 Mechanical Properties of the Specimens Fabricated with ABS Filament in X, Y, Z

For ABS filament, the effect of altering nozzle temperature and printing speed on the mechanical properties of fabricated parts in the three main directions i.e. X, Y, and Z was investigated. We did not investigate the mechanical properties in 45 degree for ABS material, as this orientation showed the weakest properties for PLA. The obtained Young's modulus, yield strength, and ultimate tensile strength of the fabricated ABS (set-B) specimens are shown in Figs. 8.8, 8.9, and 8.10. The results indicate that the part printed in X and Z orientations have better mechanical properties compared with the parts printed in Y orientation. In these samples, the highest Young's modulus achieved for the parts printed in X orientation; however, the maximum yield strength and ultimate strength achieved for the parts printed in either X or Z (however Z shows a bit better properties).

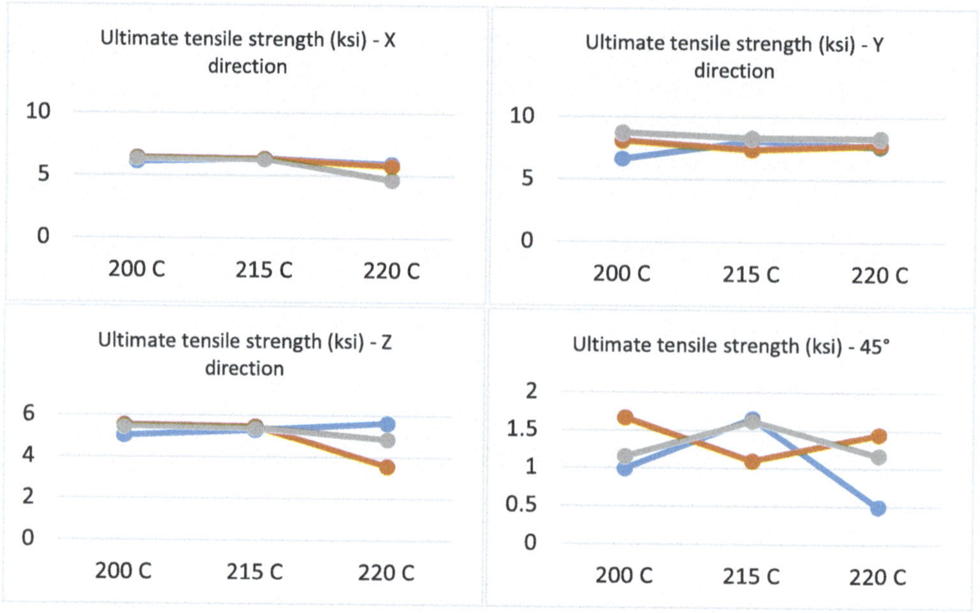

Fig. 8.7 Ultimate tensile strength in X, Y, Z orientations and 45° for the part printed with PLA in different nozzle temperatures and printing speeds

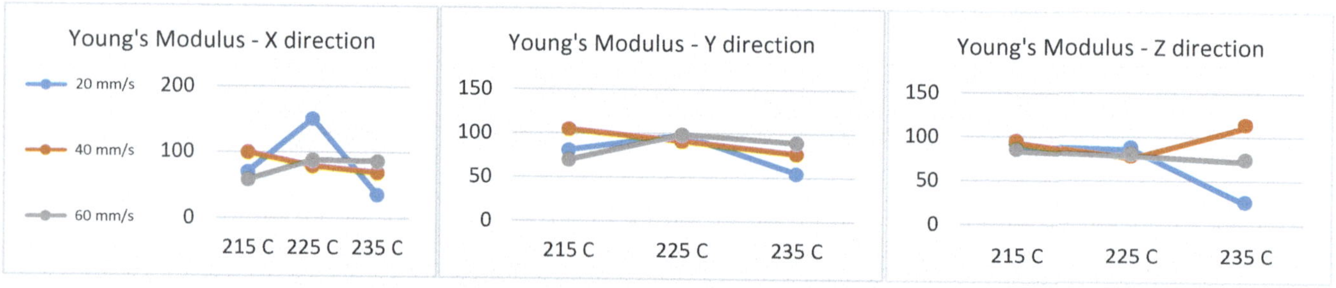

Fig. 8.8 Young's modulus in X, Y, and Z orientations for the part printed with ABS in different nozzle temperatures and printing speeds

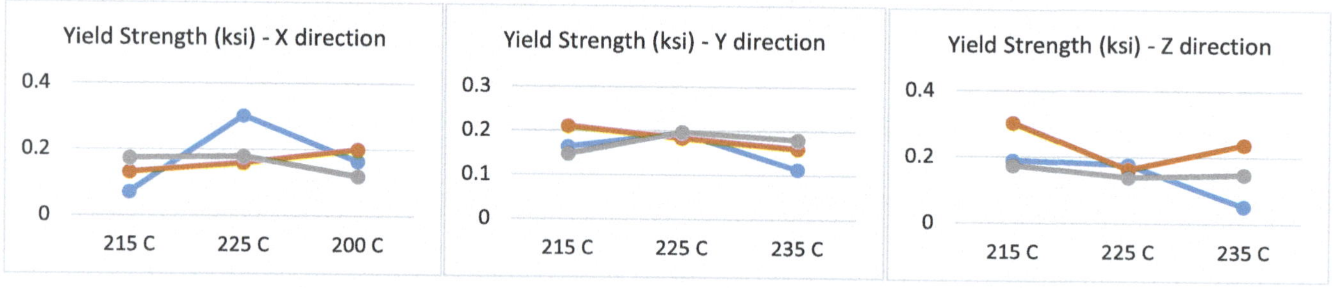

Fig. 8.9 Yield strength in X, Y, and Z orientations for the part printed with ABS in different nozzle temperatures and printing speeds

By considering of Young's modulus, X is the best orientation suggested for printing of a part with ABS. The diagrams also depict that the highest mechanical properties in X-direction were obtained at the temperature of 225 °C and the printing speed of 20 mm/s. Increasing or decreasing of the temperature will weaken the mechanical properties significantly. While the best mechanical properties for Z-direction were predominantly obtained at the temperature of 235 °C and the printing speed of 40 mm/s. However, the Yield strength shows a better result for the temperature at 215 °C and the printing speed of 20 mm/s. It is noteworthy that the weakest mechanical properties obtained for Z direction and for the temperature at 235 °C and the printing speed of 20 mm/s.

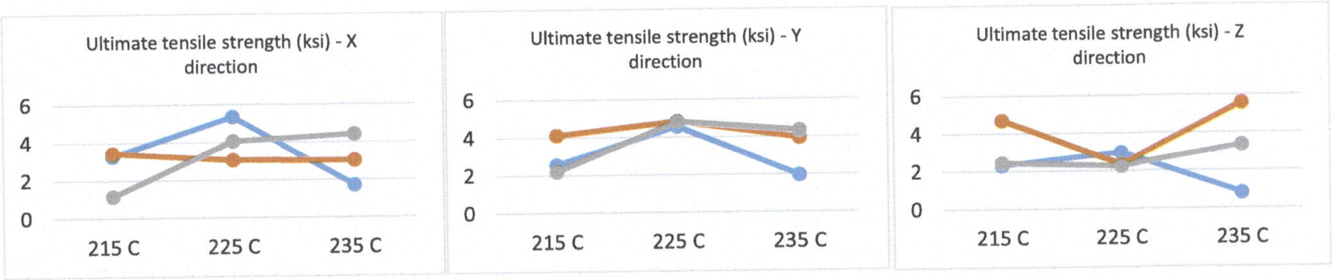

Fig. 8.10 Ultimate tensile strength in X, Y, and Z orientations for the part printed with ABS in different nozzle temperatures and printing speeds

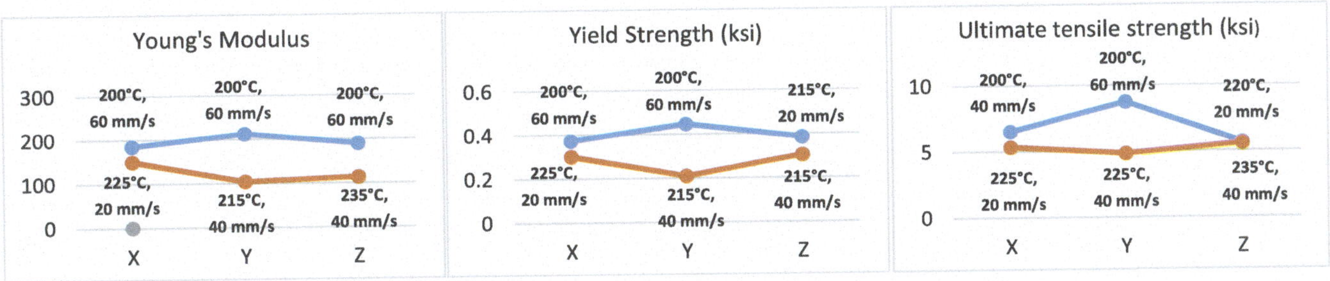

Fig. 8.11 Comparison between the maximum mechanical properties of the part printed with PLA and ABS

Figure 8.11 compares the maximum mechanical properties obtained for the specimens printed with PLA and ABS in different orientations. As this figure depicted, PLA shows stronger mechanical properties compared with ABS.

8.4 Conclusions and Future Works

In this research, the effect of printing parameters on the mechanical properties of the PLA and ABS specimens produced by fused deposition modeling process was investigated. The work reveals that PLA shows stronger mechanical properties compared with ABS. These experimental results also suggest that the specimens printed in the Y-axis orientation present the best mechanical properties in PLA while X-axis orientation is the best orientation to print parts with ABS. The optimized process parameters are recommended for the tested materials in the revealed range of parameters. More specimens need to be printed for each set of parameters to measure the experimental error. Furthermore, the obtained experimental data needs to be analyzed and explained scientifically.

References

1. E. Matias, B. Rao.: 3D printing: On its historical evolution and the implications for business. In: Management of Engineering and Technology (PICMET), 2015 Portland International Conference on, 2015, pp. 551–558
2. Dorigato, A., Moretti, V., Dul, S., Unterberger, S., Pegoretti, A.: Electrically conductive nanocomposites for fused deposition modelling. Synth. Met. **226**, 7–14 (2017)
3. Ravi, A.K., Deshpande, A., Hsu, K.H.: An in-process laser localized pre-deposition heating approach to inter-layer bond strengthening in extrusion based polymer additive manufacturing. J. Manuf. Process. **24**, 179–185 (2016)
4. Onwubolu, G.C., Rayegani, F.: Characterization and optimization of mechanical properties of ABS parts manufactured by the fused deposition modelling process. Int. J. Manuf. Eng. **2014**, 1–13 (2014)
5. A. Stephen, K. Dalgarno, J. Munguia.: Quality assurance and process monitoring of fused deposition modelling made parts. In: High Value Manufacturing: Advanced Research in Virtual and Rapid Prototyping: Proceedings of the 6th International Conference on Advanced Research in Virtual and Rapid Prototyping, 2013, pp. 31–35
6. Ahn, S.-H., Montero, M., Odell, D., Roundy, S., Wright, P.K.: Anisotropic material properties of fused deposition modeling ABS. Rapid Prototyp. J. **8**, 248–257 (2002)

Chapter 9
Mechanical Characterization of Cellulose Nanofibril Materials Made by Additive Manufacturing

Lisa M. Mariani, John M. Considine, and Kevin T. Turner

Abstract Cellulose nanomaterials have high specific stiffness and strength, are optically transparent, and are biodegradable, making them an attractive building block for bulk materials. The overall dimensions of neat bulk cellulose nanofibril (CNF) materials is significantly limited by the development of residual stresses generated during the drying process, when the source CNF is 1.0 wt.% in water or less, or by agglomeration, when the source CNF is greater than 1.0 wt.%. Here, we overcome these issues by producing CNF films and structures by additive manufacturing (i.e., 3D printing) of a shear thinning aqueous CNF suspension onto hydrophobic substrates under controlled drying conditions. Films of enhanced thicknesses, greater than 80 μm, are achieved as a result of the multistep layer-by-layer manufacturing process. The mechanical properties of the resulting materials are characterized via nanoindentation and tensile testing. Nanoindentation is used primarily to map the mechanical properties and examine variations in properties spatially and through the thickness. Tensile testing, with strain measurement via digital image correlation, is used to characterize the bulk properties. Mechanical characterization is supported by additional characterization via atomic force, optical, and electron microscopy. This study demonstrates the ability to additively manufacture stiff, strong, uniform, and scalable cellulose nanofibril materials.

Keywords Nanocellulose · Additive · DIC · Tensile · Nanoindentation

9.1 Introduction

Cellulose nanofibrils (CNFs) have low molecular weight, high aspect ratio, can be made to be optically transparent, are biodegradable, and have a variety of applications, including coatings, composites, and packaging [1–3]. Current preparation techniques for neat CNF films consist of dish drying, oven drying, and vacuum filtration techniques, which have been shown to achieve Young's moduli greater than 11 GPa and tensile strengths greater than 200 MPa [4–10]. The thickness of CNF materials with high modulus and strength has been limited to 70 μm because of large stresses that develop during drying [11].

3D printing is a technique with which thicker CNF materials may be fabricated. A few prior studies have investigated 3D printing of pure cellulose materials. CNCs, bacterial nanocelluloses, and cellulose acetate have been printed to yield structures and patterns that display shear alignment, Young's moduli up to 2.2 GPa, and tensile strengths up to 45 MPa [12–15]. For CNFs, one study printed CNF structures which maintained their intended shape using two drying techniques: freeze drying and solvent exchange [16]. Tensile strip specimens made by freeze drying had a Young's modulus and tensile strength of 0.031 ± 0.019 GPa and 3.9 ± 2.7 MPa, respectively; while, those made by air drying had a Young's modulus and tensile strength of 4.3 ± 0.3 GPa and 114 ± 14 MPa, respectively [16].

This study fabricates neat CNF materials via 3D printing and achieves films of high modulus, uniform mechanical properties, and thicknesses greater than those achieved using classical thin film preparation techniques.

L. M. Mariani (✉) · K. T. Turner
Department of Mechanical Engineering and Applied Mechanics, University of Pennsylvania, Philadelphia, PA, USA
e-mail: lmariani@seas.upenn.edu

J. M. Considine
USDA Forest Service, Forest Products Laboratory, Madison, WI, USA

S. Kramer et al. (eds.), *Mechanics of Additive and Advanced Manufacturing, Volume 8*, Conference Proceedings of the Society for Experimental Mechanics Series, https://doi.org/10.1007/978-3-319-95083-9_9

9.2 Methods

Aqueous suspensions of 0.93 wt.% TEMPO pretreated CNFs are printed on a commercial 3D printer (i3PRO, MakerFront) that was modified to have a syringe print head with controlled back pressure. CNF suspensions were printed in geometries of single line/single layer, single line/multiple layers, and 6 line passes/2 layers (bi-layer specimens) for tensile testing and were subsequently cured by controlling the substrate temperature. Suspensions were printed on hydrophobized glass and silicon build plates with a constant back pressure of 172 kPa and with varying print head velocities and substrate temperatures. The surface and shape of the 3D printed CNFs are imaged using mechanical profilometry (P7, KLA-Tencor), optical microscopy (Axio, Zeiss), tapping mode atomic force microscopy (AFM) (Dimension 2000 AFM, Bruker), and electron microscopy (7500F, Jeol).

Mechanical properties are characterized using nanoindentation and tensile testing. Nanoindentation (TI 950 Triboindenter, Hysitron) measurements of 70 °C substrate temperature single line/single layer prints on silicon substrates were performed to evaluate variations of the reduced modulus in the 3D printed CNF materials. Arrays of nanoindentation indents were set up along the length of the line and width of the line with a minimum indent spacing of 50 μm. Both a Berkovich diamond tip and a spherical tip with a radius of 0.814 μm were used for nanoindentation to determine if the fiber specimens could be treated as homogenous materials. The Young's modulus and tensile strength of the CNFs were investigated using tensile testing. Bi-layer strip specimens marked with glitter were loaded and strained in a uniaxial testing machine (Criterion Model 43, MTS) equipped with a 50 N load cell (LSB.501, MTS) and digital camera (Guppy Pro, Allied Vision) fitted a manual focus lens (HR F2.8/50 mm, Navitar) for digital image correlation. The force-displacement data was fit with a hyperbolic tangent expression. From the derivative of this expression, the specimen cross-sectional area, and the gauge length, the Young's modulus, E, was determined by evaluating the stiffness of the specimen at zero displacement. The density of the printed specimens was determined by measuring the volume using digital calipers and a digital micrometer and the weight of the specimen using a precision balance (Adventurer SL, Ohaus).

9.3 Results and Discussion

Mechanical profilometry measurements of the line width, line height, and cross sectional area of the single line/single layer prints show that the line width and line height follow a decreasing power-law trend with increasing print head speed and the cross-sectional area of the line is independent of substrate temperature. The rms roughness of the lines was probed with AFM and was measured to be 3.63 nm ± 0.21 nm. Additionally, the single line/multiple layer prints achieved thicknesses greater than 80 μm.

The uniformity of the mechanical properties of single line/single layer CNF prints was revealed through nanoindentation measurements, where there was no significant variation in the reduced modulus along the length or width of the line. The reduced moduli of the CNF prints measured by the Berkovich and spherical indenters were 10.5 ± 0.41 GPa and 10.3 ± 0.44 GPa, respectively, which suggests that the surface of the CNF films can be assumed to be homogenous. The hardness was measured to be 0.67 GPa ± 0.03 GPa. Tensile testing with DIC revealed a Young's modulus $E = 10.2$ GPa ± 1.2 GPa. The films also had a tensile strength at fracture of $\sigma_f = 72.6$ MPa ± 7.4 MPa. The specific stiffness and specific strength of the 3D printed films was 8048 ± 1384.3 kN-m/kg and 57.3 ± 8.3 kN-m/kg, respectively, based on a measured density of 1267.4 ± 131.1 kg/m^3.

9.4 Conclusions

Cellulose nanofibril materials with uniform mechanical properties are prepared using additive manufacturing. The 3D printed CNFs have high modulus, $E = 10.2$ GPa ± 1.2 GPa, and moderate strength, $\sigma_f = 72.6$ MPa ± 7.4 MPa. Additionally, this study exceeds 80 μm thicknesses for multi-layer line prints, which pushes the boundaries of the current thicknesses achieved for CNF materials.

References

1. Shatkin, J.A., Wegner, T.H., Bilek, E.M., Cowie, J.: Market projections of cellulose nanomaterial-enabled products – part 1: applications. Nanocellulose Mark. **13**(5), 9–16 (2014)
2. Isogai, A.: Wood nanocelluloses: fundamentals and applications as new bio-based nanomaterials. J. Wood Sci. **59**(6), 449–459 (Sep. 2013)
3. Abdul Khalil, H.P.S., Bhat, A.H., Ireana Yusra, A.F.: Green composites from sustainable cellulose nanofibrils: a review. Carbohydr. Polym. **87** (2), 963–979 (Jan. 2012)
4. Aulin, C., Salazar-Alvarez, G., Lindström, T.: High strength, flexible and transparent nanofibrillated cellulose–nanoclay biohybrid films with tunable oxygen and water vapor permeability. Nanoscale. **4**(20), 6622 (2012)
5. Malho, J.M., Laaksonen, P., Walther, A., Ikkala, O., Linder, M.B.: Facile method for stiff, tough, and strong nanocomposites by direct exfoliation of multilayered graphene into native nanocellulose matrix. Biomacromolecules. **13**(4), 1093–1099 (2012)
6. Fukuzumi, H., Saito, T., Iwata, T., Kumamoto, Y., Isogai, A.: Transparent and high gas barrier films of cellulose nanofibers prepared by TEMPO-mediated oxidation. Biomacromolecules. 162–165 (2009)
7. Sehaqui, H., Ezekiel Mushi, N., Morimune, S., Salajkova, M., Nishino, T., Berglund, L.A.: Cellulose nanofiber orientation in nanopaper and nanocomposites by cold drawing. ACS Appl. Mater. Interfaces. **4**(2), 1043–1049 (2012)
8. Wu, C.N., Saito, T., Fujisawa, S., Fukuzumi, H., Isogai, A.: Ultrastrong and high gas-barrier nanocellulose/clay-layered composites. Biomacromolecules. **13**(6), 1927–1932 (2012)
9. Zhu, H., Zhu, S., Jia, Z., Parvinian, S., Li, Y., Vaaland, O., Hu, L., Li, T.: Anomalous scaling law of strength and toughness of cellulose nanopaper. Proc. Natl. Acad. Sci. **112**(29), 8971–8976 (2015)
10. Henriksson, M., Berglund, L.A., Isaksson, P., Lindstrom, T., Nishino, T.: Cellulose Nanopaper structures of high toughness. Biomacromolecules. **9**(6), 1579–1585 (2008)
11. Baez, C., Considine, J., Rowlands, R.: Influence of drying restraint on physical and mechanical properties of nanofibrillated cellulose films. Cellulose. **21**(1), 347–356 (Jan. 2014)
12. Marksteadt, K., Sundberg, J., Gatenholm, P.: 3D bioprinting of cellulose structures from an ionic liquid. 3D Print. Addit. Manuf. **1**(3), 115–121 (2014)
13. Pattinson, S.W., Hart, A.J.: Additive manufacturing of cellulosic materials with robust mechanics and antimicrobial functionality. Adv. Mater. Technol. **2**, 1600084 (2017)
14. Siqueira, G., Kokkinis, D., Libanori, R., Hausmann, M.K., Gladman, A.S., Neels, A., Tingaut, P., Zimmermann, T., Lewis, J.A., Studart, A.R.: Cellulose nanocrystal inks for 3D printing of textured cellular architectures. Adv. Funct. Mater. **27**(12), 1604619 (2017)
15. Rees, A., Powell, L.C., Chinga-Carrasco, G., Gethin, D.T., Syverud, K., Hill, K.E., Thomas, D.W.: 3D bioprinting of carboxymethylated-periodate oxidized nanocellulose constructs for wound dressing applications. Biomed Res. Int. **2015**, 925757 (2014)
16. Hakansson, K.M.O., Henriksson, I.C., Pena Vazquez, C., Kuzmenko, V., Markstedt, K., Enoksson, P., Gatenholm, P.: Solidification of 3D printed nanofibril hydrogels into functional 3D cellulose structures. Adv. Mater. Technol. **1**, 1600096 (2016)

Chapter 10
Shock Propagation and Deformation of Additively-Manufactured Polymer Foams with Engineered Porosity

Jonathan E. Spowart, David Lacina, Christopher (Kit) Neel, Geoffrey Frank, Andrew Abbott, and Brittany Branch

Abstract The propagation of shocks through additively-manufactured (AM) polymeric structures containing multiple length scales of engineered porosity is studied both experimentally and computationally. In this study, a single-stage light gas gun is used to impact cube-shaped specimens, 40 mm on a side, instrumented with photon Doppler velocimeter (PDV) light probes to capture free surface velocities and side-looking high-speed video to capture deformation history. A combined Eulerian-Lagrangian finite element (FE) model has been developed which reproduces the majority of the observed experimental trends, based on an independently-measured shock Hugoniot for the bulk AM polymer. After initial calibration, the FE model has been used to suggest candidate geometries for experimental investigation, based on the desired shock response. Geometries for structurally-efficient shock mitigation have been investigated. In a separate set of experiments, miniature (6 mm × 6 mm) square specimens have been impacted at the Dynamic Compression Sector at the Advanced Photon Source (APS), and imaged using x-ray Phase Contrast Imaging (PCI). This technique gives strong evidence for the propagation of discrete shocks within the engineered foam structures, in agreement with our models.

Keywords Shock · Porous · Fractal · Additive manufacturing · Polymer

10.1 Introduction

Dissipative phenomena in nature often exhibit hierarchical or multi-scale structures. For example, consider the three structures in Fig. 10.1 below, which shows the responses of different materials corresponding to the dissipation of (a) volumetric strain energy – the desiccation cracking of mud [1]; (b) electrical potential energy – a Lichtenberg figure [2], and (c) elastic strain energy – small-caliber bullet penetrating a glass sheet [3]. In each case, the resulting structure comprises multiple length scales, arranged in a hierarchical fashion. This is so much the case that it is often unnecessary to include scale bars on these figures, since the structures repeat at finer and finer scales – this is the essence of *fractal* [4] geometry.

The example of the Lichtenberg figure is particularly representative in that dissipation of electrical energy is clearly from top to bottom in the figure. As the energy is dissipated, more and more branches are created that are at finer and finer length scales. It is specifically this kind of geometry that we hypothesize would be useful for shock wave mitigation if engineered directly into the material itself. The 3-D fractal geometry that we chose to investigate is called a Menger sponge [5, 6], examples of which are given below in Fig. 10.2. This mathematically defined hierarchical structure is developed iteratively by removing successive prismatic volumes from a solid cube, where each iteration is 1/3 of the size of the previous one, until, in the mathematical limit, the structure has zero solid volume and infinite surface area. Figure 10.2 shows the 1st, 2nd, and 3rd

J. E. Spowart (✉)
Air Force Research Laboratory, Materials and Manufacturing Directorate, AFRL/RXCCM, Wright-Patterson AFB, OH, USA
e-mail: jonathan.spowart@us.af.mil

D. Lacina
University of Dayton Research Institute, Eglin AFB, FL, USA

C. Neel
Air Force Research Laboratory, Munitions Directorate, AFRL/RWMWS, Eglin AFB, FL, USA

G. Frank · A. Abbott
University of Dayton Research Institute, Dayton, OH, USA

B. Branch
Los Alamos National Laboratory, Nuclear Materials Science (MST-16), Los Alamos, NM, USA

S. Kramer et al. (eds.), *Mechanics of Additive and Advanced Manufacturing, Volume 8*, Conference Proceedings of the Society for Experimental Mechanics Series, https://doi.org/10.1007/978-3-319-95083-9_10

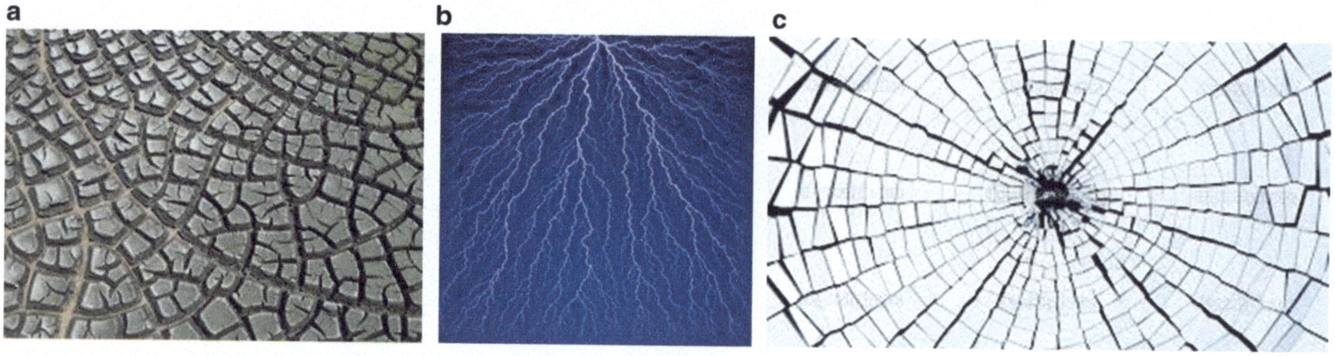

Fig. 10.1 Showing different material responses occurring during the dissipation of (**a**) volumetric strain energy (desiccation cracking in mud) [1], (**b**) electrical potential energy (Lichtenberg figure) [2], and (**c**) elastic tensile strain energy (small caliber bullet penetrating glass sheet) [3]

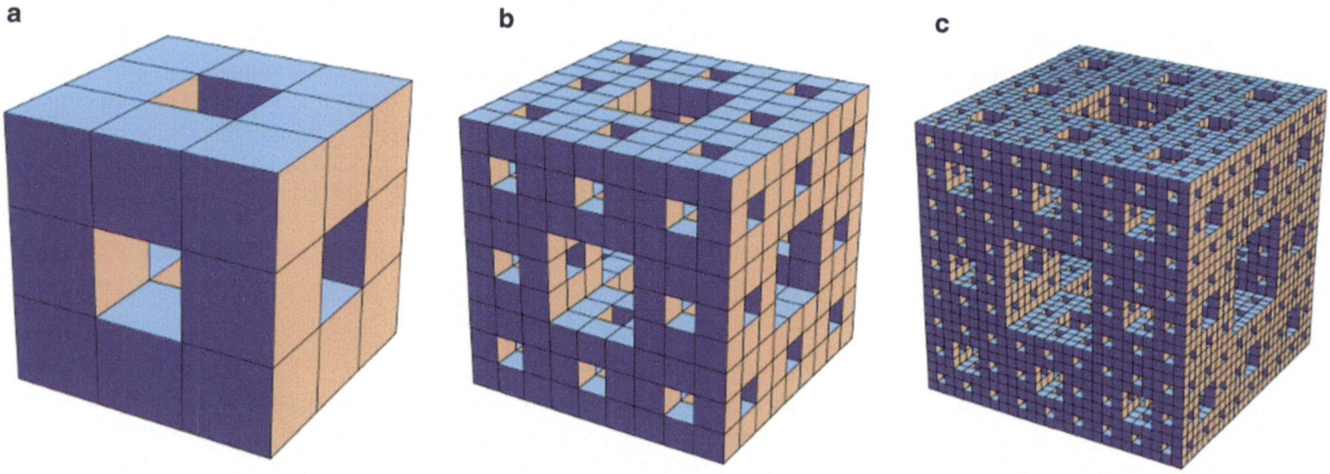

Fig. 10.2 Showing the (**a**) 1st, (**b**) 2nd, and (**c**) 3rd iterations of the Menger sponge [5, 6], with solid volume fractions of 0.741, 0.549, and 0.406. In the mathematical limit, the volume fraction is zero and the surface area is infinite

iterations of the Menger sponge, with solid volume fractions of 0.741, 0.549, and 0.406 respectively. In general, the variation of solid volume fraction is given by $v = (20/27)^i$ where $i =$ iteration number.

It is important to note that Menger sponges are geometrically different from typical lattice-based foams [7] in that they have a different mathematical relationship between their surface area (S) and volume (V). Figure 10.3 below, shows the variation of the surface area to volume ratio, $\xi = S/V$, as a function of specific volume ($1/\rho$) for a Menger sponge, compared to the same parameter calculated for two different closed-cell foams based on space-filling tetrakaidecahedra.

Significant differences between the fractal and lattice-based foams, especially at high and low specific volumes, suggest that the compressive deformation of these hierarchical structures (including their pore collapse) will follow different dynamics than lattice-based structures, resulting in differences in their shock propagation behavior and therefore providing motivation for the current study.

10.2 Experimental

10.2.1 Additive Manufacturing

For the plate impact experiments, we chose to print 2-D versions of the Menger sponges, where prismatic volumes were successively removed only from one face which contained the impact direction. In this way, we could confine the experiment

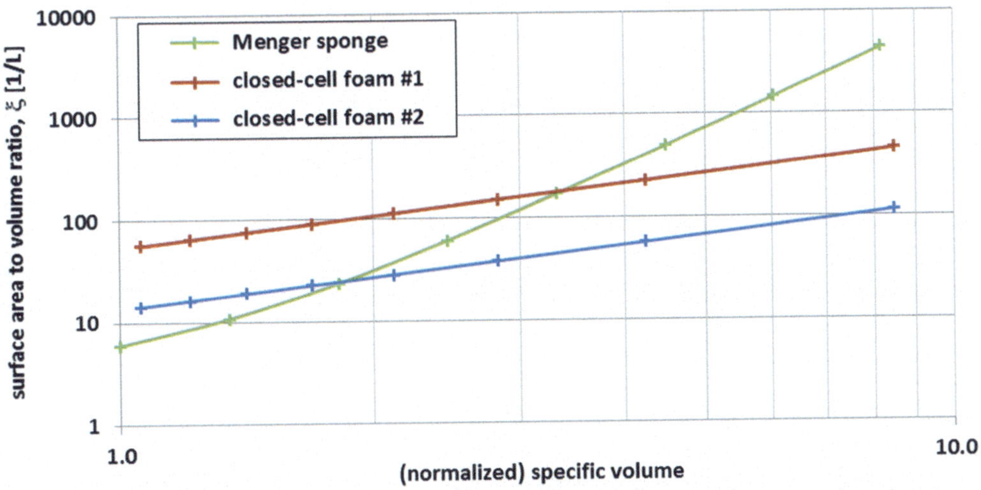

Fig. 10.3 Relationship between the parameter ξ, and normalized specific volume, *V*, for Menger sponge (green), and two different closed-cell foams [7] based on space-filling tetrakaidecahedra (red and blue). The slopes of the lines (dξ/d*V*) suggest that the fractal foam should behave very differently under dynamic compression, compared to the closed-cell foams, which should both behave similarly

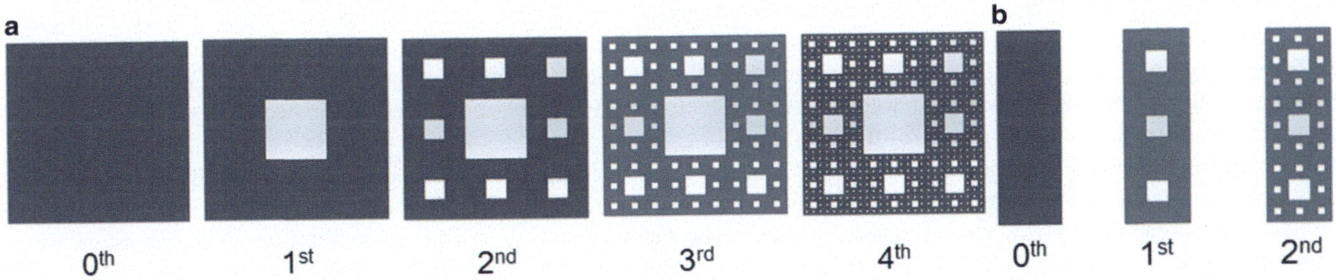

Fig. 10.4 (**a**) 40 mm cube Menger geometries ranging in fractal iteration order from 0th to 4th, and (**b**) 13.33 mm width specimens used for some experiments. Note the 2nd and 3rd iterations in (**a**) in part of the 1st and 2nd iterations in (**b**)

to mainly planar strain conditions, which greatly simplified both the numerical modeling, and interpretation of the high-speed video of the deformation. These 2-D structures are called "prismatic" sponges to differentiate from the 3-D Menger sponges shown above. For 2-D prismatic sponges, the variation of solid volume (v) with each iteration is given by $v = (8/9)^i$ where $i =$ iteration number. 2-D sponge geometries were additively manufacturing utilizing two different techniques; both with photopolymer resins. The first additive manufacturing technique used was polyjet performed by a Stratasys Objet260 Connex 3. This printer jets photopolymer resin from several small orifices onto a substrate or previous layer and quickly cures the resin with UV light. Digital ABS material was selected as the model material for its ABS-like mechanical properties and plasticity. This material is produced by combining two proprietary photopolymer resins designated RGD515 and RGD531, with a matte surface finish.

In addition, two-part LANL wedge specimens [8] were printed for shock Hugoniot measurements of the digital ABS material. Details of these measurements are included in the next section, and elsewhere [9]. A number of different 2-D prismatic Menger sponge geometries were printed on the Stratasys Objet260 Connex 3. The first was a cubic geometry measuring 40 mm on each side. Five fractal iterations were also printed ranging from 1st-order to 4th-order as shown in Fig. 10.4a. The practical resolution limit of this printer is around 200 μm. The 0th-order (cube) served as a reference specimen for shock wave traversal time in the solid material, and for and calibrating the material's spall strength – see "Modeling" below. A second geometry measured 40 mm in length and depth, but only 13.33 mm (40/3 mm) in width, in order to avoid the influence of the specimen edges on the back-surface motion during the experiment [11]. Two fractal iterations were printed using the 13.33 mm geometry; 1st-order and 2nd-order (Fig. 10.4b). The 0th-order (solid) was omitted because shock speeds were not expected to differ significantly from those in the 40 mm cube.

The second additive manufacturing technique utilized in specimen fabrication was digital light projection (DLP) – a form of stereolithography (SLA). DLP printing was performed with an Autodesk Ember printer. In this technique, a UV light reflects off a digital micromirror which orients to produce a 2D image of the current layer being printed. The image is projected into a photopolymer resin bath, curing the resin. In this way, an entire layer is printed at once. The build head then lifts up by one layer height creating a gap filled by fresh resin for the next layer. The printer can maintain a high resolution of 50 μm in-plane and 10 μm out-of-plane. The high resolution of the printer enabled the printing of prismatic Menger geometries measuring 6 mm × 2 mm × 6 mm required for the x-ray phase contrast imaging (PCI) experiments. These were printed with standard clear PR48 acrylate-based resin from Colorado Photopolymer Solutions. Fractal geometries of the 1st and 2nd iterations were printed as shown in Fig. 10.4b. The printer was able to resolve 2nd order channels measuring 222 μm, albeit somewhat inconsistently. Surface tension and capillary forces combined with scattering of UV light caused curing of resin in some of these channels. A large batch of specimens was printed to obtain enough specimens with clear channels for high rate impact experimentation.

10.2.2 High Rate Impact Testing

10.2.2.1 Plate Impact Testing at AFRL/RWMW's HP3 Facility

Two different plate impact experiments were carried out at AFRL/RWMW's HP3 gun impact facility, in order to (i) measure the solid printed ABS material's shock Hugoniot and obtain an accurate U_s–u_p relationship for use in FEA models, and (ii) to obtain high speed video of the shock deformation of specimens with different levels of engineered porosity, comparing fractal and regular grid geometries. All Hugoniot experiments were performed using a symmetric impact configuration. A 60 mm single-stage powder gun and 4″ gas gun (wrap around breech) were used to accelerate printed ABS impactors to high velocities, producing nominally planar shockwaves in printed ABS double-wedge [8] type targets upon impact. Embedded EM gauges (of the LANL type) [9] were used to measure particle velocities (u_p) and shock velocities (U_s) as a function of depth within the specimens. In addition, photon Doppler velocimetry (PDV) was employed on the back face of each specimen in order to measure the particle velocity of this surface.

For the second set of experiments, it was hoped that the high speed video could be used to show qualitative agreement between the experiments and the models. For quantitative comparisons, photon Doppler velocimetry (PDV) was employed on the back face of each specimen, and compared with nodal displacements *vs.* time from the simulations. The specimens were sputter coated with a thin layer of Al to aid with reflectivity during the PDV experiments. Video was obtained at 5,000,000 frames per second, using a Xenon flash lamp for illumination. An exposure time of 100 ns was sufficient to avoid blurring of the images. Figure 10.5 below shows the experimental set-up.

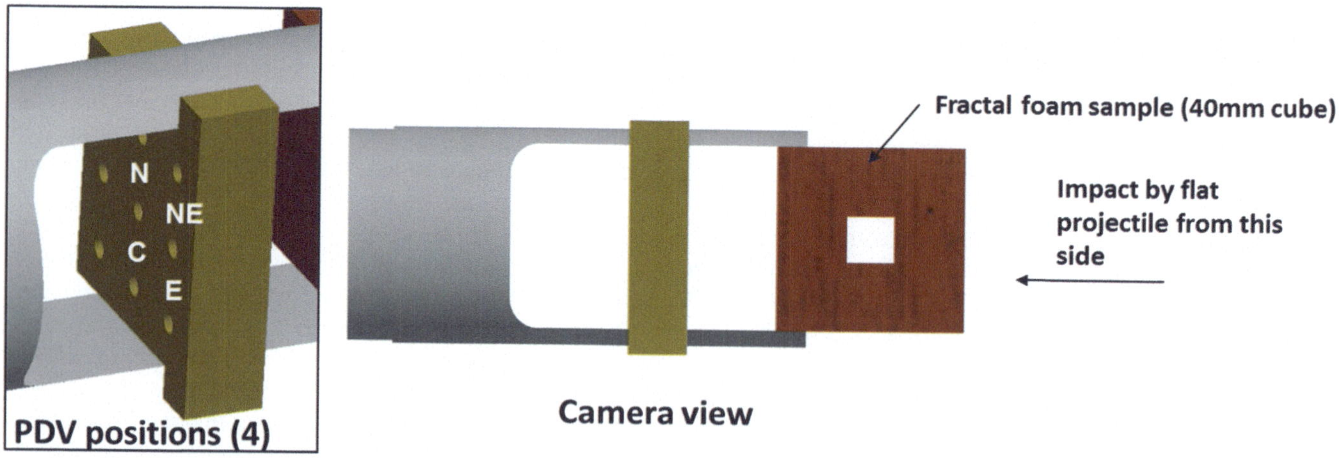

Fig. 10.5 Schematic of the experimental configuration for plate impact experiments of engineered foam specimens (40 mm cubes and 40 mm × 40 mm × 13.3 mm slabs) incorporating high-speed video at 5,000,000 frames per second and four PDV probes (N, NE, C, E) on the back surface of the specimen. The specimen is minimally clamped in the fixture to minimize edge effects from the clamps

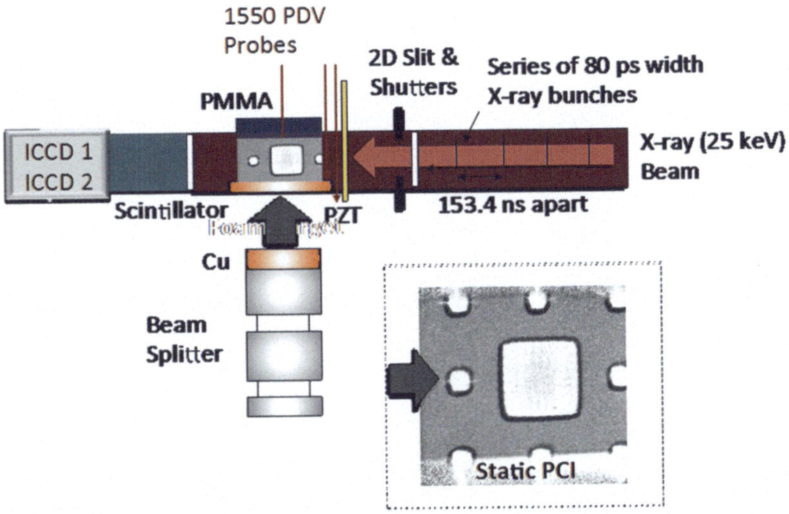

Fig. 10.6 Schematic of the experimental configuration for transmission PCI. The gun is oriented perpendicular to a series of 80-ps width X-ray bunches (E = 25 ± 0.9 keV, λ = 0.5386 Å) spaced 153.4 ns apart transmitted through the AM polymer foam (inset) and detected using a LuAg:Ce ($Lu_3Al_5O_{12}$:Ce) scintillator optically coupled to four independent image intensified charge coupled device (ICCD) detectors (Princeton Instruments) to provide 4 to 8 X-ray images per experiment

10.2.2.2 Phase-Contrast Imaging Techniques at Advanced Photon Source

Shock compression of AM polymer foams combined with real-time phase contrast imaging (PCI) and photon Doppler velocimetry (PDV) were examined by impact experiments conducted at the Dynamic Compression Sector (DCS) using the IMPact system for the ULtrafast Synchrotron Experiments (IMPULSE). This unique capability at the Advanced Photon Source (APS) allowed a more detailed understanding of the wave dynamics as a result of porosity. The foam (2 mm × 6 mm × 6 mm) was epoxied (Angstrom Bond) to a PMMA window (3 mm × 6 mm × 6 mm) with a 0.8 μm Al mirror deposited on the foam/PMMA interface. A 1.0 mm thick Cu (OFHC-Cu) impact plate was affixed to the opposing surface of the micro-lattice sample to act as a drive plate and the assembly was secured to a target holder. Polycarbonate projectiles faced with Cu impactors were accelerated at 0.303 km/s using a single stage powder gun resulting in a pressure of 0.98 GPa [10]. Figure 10.6 shows a schematic of the experimental configuration for transmission PCI with the gun oriented perpendicular to a series of 80-ps width X-ray bunches (E = 25 ± 0.9 keV, λ = 0.5386 Å) spaced 153.4 ns apart transmitted through the AM polymer foam (inset) and detected using a LuAg:Ce ($Lu_3Al_5O_{12}$:Ce) scintillator optically coupled to four independent image intensified charge coupled device (ICCD) detectors (Princeton Instruments) to provide 4 to 8 X-ray images per experiment [11–13].

10.3 Modeling

Finite element modeling was performed using Abaqus/Explicit. Models have multiple contacting bodies, with flyer plates and anvils modeled as Lagrangian regions and a target lying in between the flyer and anvil modeled as an Eulerian region. Modeling practices followed standard Abaqus guidelines [14]. First-order, reduced integration hexahedral elements are used in Lagrangian regions. First-order multi-material, reduced integration hexahedral elements with hourglass control are used in Eulerian regions. Second-order advection is used in all Eulerian regions. All materials are modeled as isotropic. Boundary conditions representative of geometric symmetry are used wherever appropriate. Eulerian-Lagrangian contact is used at part-to-part interfaces, with a fiction coefficient of 1.0.

10.3.1 Constitutive Model

Constitutive models for materials used in the FEA utilize an elastic-plastic model for deviatoric response and a Mie-Grüneisen equation of state (EOS) for dilatation response. Parameters for the material models are listed in Table 10.1. For the printed

Table 10.1 Parameters used for constitutive models in analysis [15–19]

Property	Units	Printed ABS	PMMA	7075-T6 Aluminum
Density	gm/cm^3	1.173	1.190	2.804
Elastic shear modulus	MPa	1350	2200	26,700
Yield strength	MPa	105	290	420
Spall strength	MPa	200	–	–
EOS – reference sound speed, c_0	m/s	2400	2210	5200
EOS – slope, s	–	1.81	1.82	1.36
EOS – Grüneisen ratio, γ_0	–	1	0.85	2.2

acrylonitrile butadiene styrene (ABS) material, the EOS parameters were developed based on the plate impact testing at AFRL/RWMW [15] and were found to be reasonably-well represented by equation [1] below:

$$U_s = 2.40 + 1.81u_p \tag{10.1}$$

Shear modulus and yield strength were estimated based on values in the literature [16, 17]. Spall strength was selected to match the predicted back-face velocity predicted by models with the velocity measured in testing of solid cubes. For the poly (methyl methacrylate) (PMMA) material the EOS parameters were adapted to the linear U_s–u_p relationship used in the Mie-Grüneisen implementation in Abaqus based on quadratic model parameters from Ref. [18]. The shear modulus and yield strength were extrapolated to 5000/second rate based on response from lower rate testing [19]. For the 7075-T6 aluminum the parameters are all based on data from Ref. [18].

10.4 Results

10.4.1 Correlation Between AFRL/RWMW Impact Testing and FEA Modeling

Printed ABS cubes, 40 mm on each side, were normally impacted at the AFRL/RWMW impact facility at nominally 500 m/s by a PMMA disk epoxied into the front of an aluminum projectile. Dimensions for the test and for the model are shown in Figs. 10.5 and 10.7a, respectively. Both video, captured at 5,000,000 frames per second with 100 ns exposure time per frame, and velocity of the back face at four locations were measured. Back-face velocity data were acquired at the four locations shown in Fig. 10.5 using photon Doppler velocimetry (PDV). Tests were conducted on the four configurations shown in Fig. 10.7b.

The measured PDV data were used for correlating with FEA. The FEA model used the procedures described above. Element sizes were on the order of 0.5 mm. Symmetry conditions were used to reduce the model to ¼ of the geometry shown in Fig. 10.7a. Because the targets are modeled as Eulerian regions, the back-face location is not well-defined. To determine back-face velocity from the FEA model, output was requested for lines of nodes parallel to the X-axis at the appropriate Y- and Z-coordinates of each PDV location. Velocities were determined from based on the most-positive nodal location which had a nonzero velocity.

Measured velocities for the solid cube and 3rd order Menger configuration are shown in Fig. 10.7c, e, respectively. FEA predicted velocities for the same two configurations are shown in Fig. 10.7d, f. Good agreement is observed between the predictions and the measured data. The data indicate that a true shock is transmitted by the solid cube, while a ramp wave reaches the back face for the 3rd order Menger cube. Not shown are results for the 1st–order Menger cube, for which the PDV traces indicate a shock reaches the back face with lower amplitude than that for the solid cube, and results for the 2nd–order Menger cube, for which a ramp wave is observed at the back face with a steeper slope than observed for the 3rd–order Menger cube. In both cases, the model agrees well with the measured data.

10.4.2 Phase Contrast Imaging (PCI) Results

The figure below (Fig. 10.8) shows the results of a representative impact experiment carried out at the Dynamic Compression Sector (DCS) at the Advanced Photon Source (APS), under PCI conditions. A copper flyer plate impacts the 2nd-order fractal

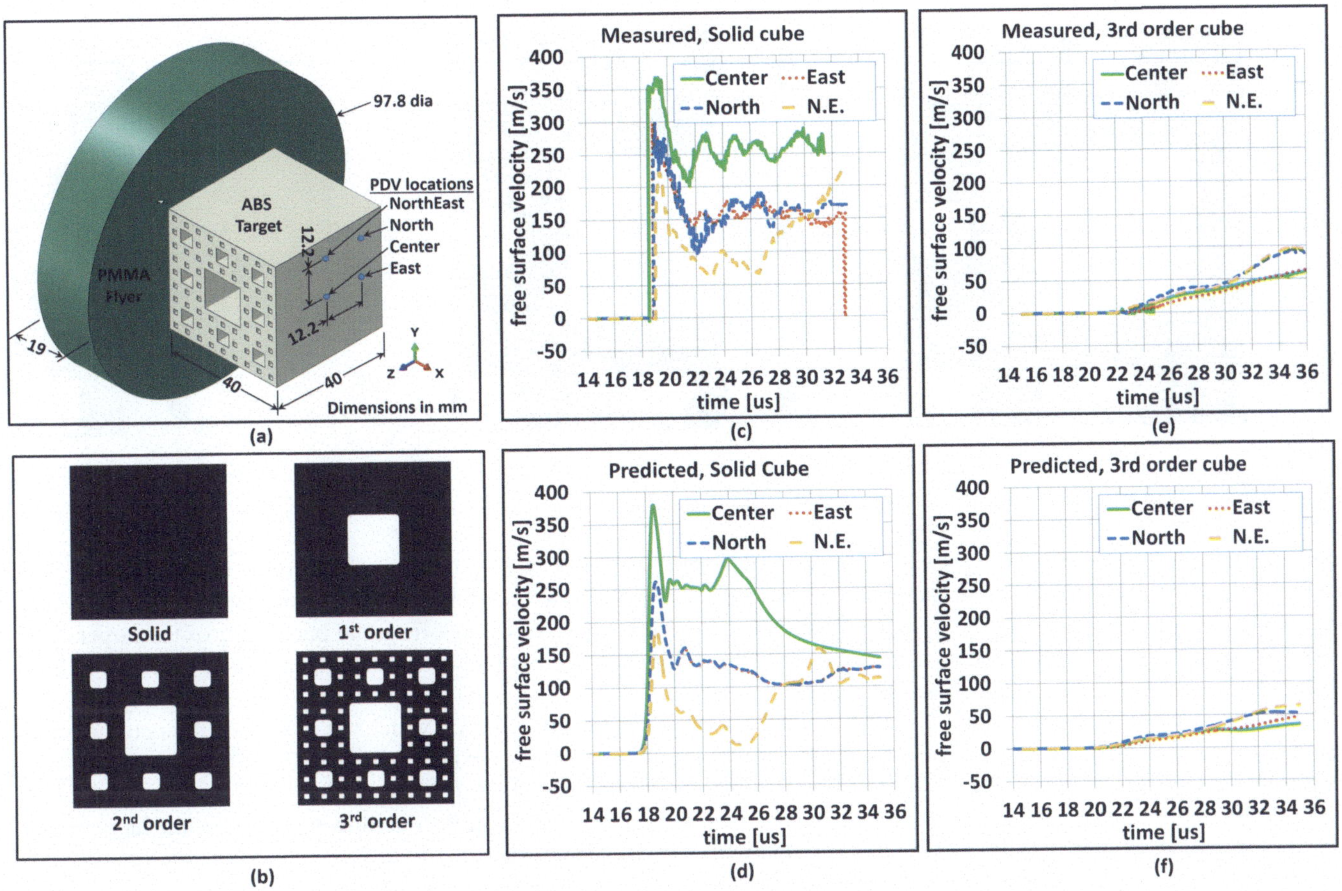

Fig. 10.7 Correlation of measurements and models for 40 mm cubes, (**a**) test configuration, (**b**) section view of four items tested, (**c**) measured back-face velocity for solid cube, (**d**) predicted back-face velocity for solid cube, (**e**) measured back-face velocity for cube of 3rd –order shape, (**f**) predicted back-face velocity for cube of 3rd –order shape. Time is measured from impact

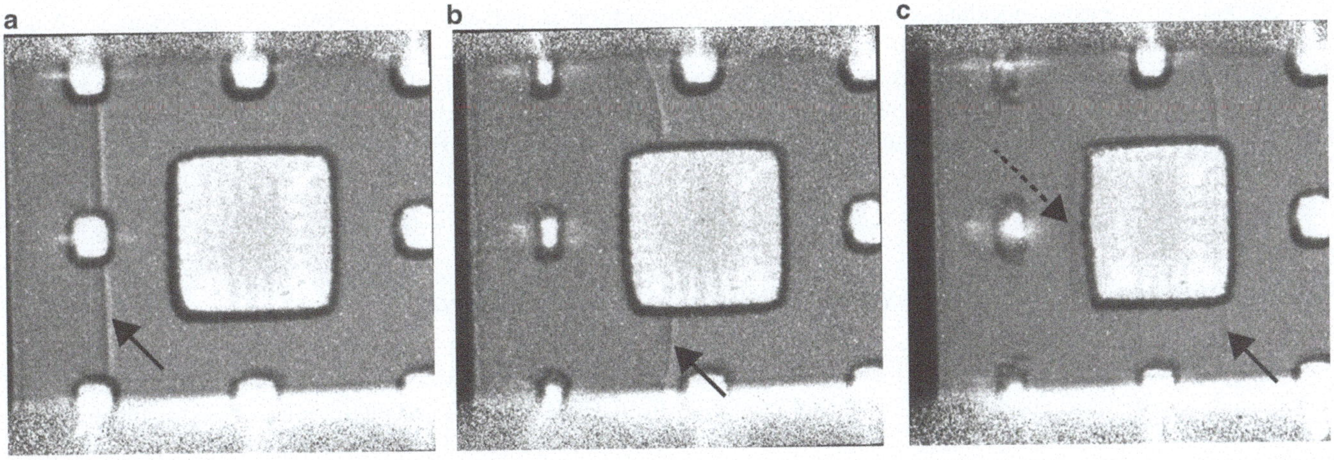

Fig. 10.8 PCI images of shock waves transiting a 2nd – order fractal geometry specimen under impact conditions. The shock wave (arrowed) is transiting from left to right in the figure. The shock wave becomes curved as it interacts with the free surface of the larger void feature. The dotted arrow is a release wave from the back surface of the central pore. The gray areas to the top and bottom are artifacts of the beamline shutter system. The field of view is 1.7 mm squared

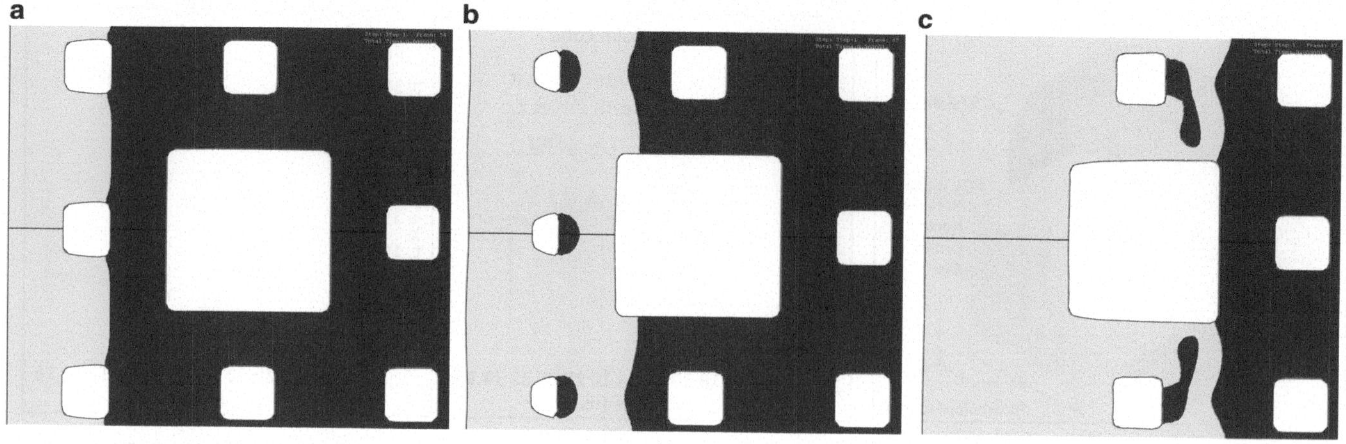

Fig. 10.9 Simulations of the PCI experiment carried out at APS. Contours delineate regions of non-zero velocity (lab reference frame) at various time steps in the simulation corresponding to Fig. 10.8a–c

specimen from left to right in the figure, at 0.303 km/s. The field of view is 1.7 mm square. Each frame is a subsequent time step (153.4 μs), showing (a) the propagation of a planar shock wave through the smaller voids, (b) curvature of the shock front as it interacts with the free edges of the larger void, and (c) exit of the curved shock from the back edge of the larger void. Previous work has shown shock modulation in lattice foam structures [20], but this is the first experimental evidence of local planar shocks propagating in fractal foam structures.

Similar structures were observed in 2-D FEA models, as shown in Fig. 10.9a–c, on which are plotted contours of non-zero velocity (lab reference frame) for various time steps during the simulated impact experiment. However there are a number of important differences, including the dynamics of pore collapse and the generation of release waves from the rearward-facing surfaces of the pores. Release waves are not generally seen in the experiments with the possible exception of the large central pore in Fig. 10.8c annotated with a dotted arrow. Quantitative discrepancies between model and experiment are most likely due to inadequacies of the material model, including viscosity and other non-linear effects.

10.5 Discussion

The reasonable correlation between the FEA model and the experimental results for the 40 mm cube specimens provided justification for an extensive modeling campaign to explore different fractal and non-fractal (regular) geometries under planar impact, both in the 40 mm and 13.3 mm specimen thicknesses. It was found to be far more efficient to investigate these different geometries via modeling than by experiment, choosing to experimentally validate only the more promising cases. Over 100 different geometries, with various solid volume fractions and different arrangements of pores were investigated in this way, which would have been highly impractical if carried out experimentally. The following section highlights some of the more interesting results from these modeling runs.

10.5.1 Variation of Anvil Contact Pressure with Specimen Geometry

To assess the ability of these engineered foam geometries to attenuate shocks, a series of analyses was performed using the FEA procedures described previously with material properties from Table 10.1. The configuration assessed is illustrated in Fig. 10.10a. A PMMA flyer impacts the ABS target, which is in direct contact with a thin layer of PMMA that is backed by a stiff object. A single layer of elements in the Z-direction is modeled, with Z-direction constraints imposing plane strain conditions. Symmetry constraints are imposed on Y-normal sides of all parts. These constraints result in the model representing a flyer, target, and anvil of infinite dimension in the Y- and Z-directions, with finite thickness in the X-direction. The flyer and anvil are modeled as Lagrangian regions, with an element size of 0.2 mm. The target is modeled as an Eulerian region, with element sizes ranging from 0.2 to 0.1 mm, depending on the size of features in the voids. Frictionless contact is assumed.

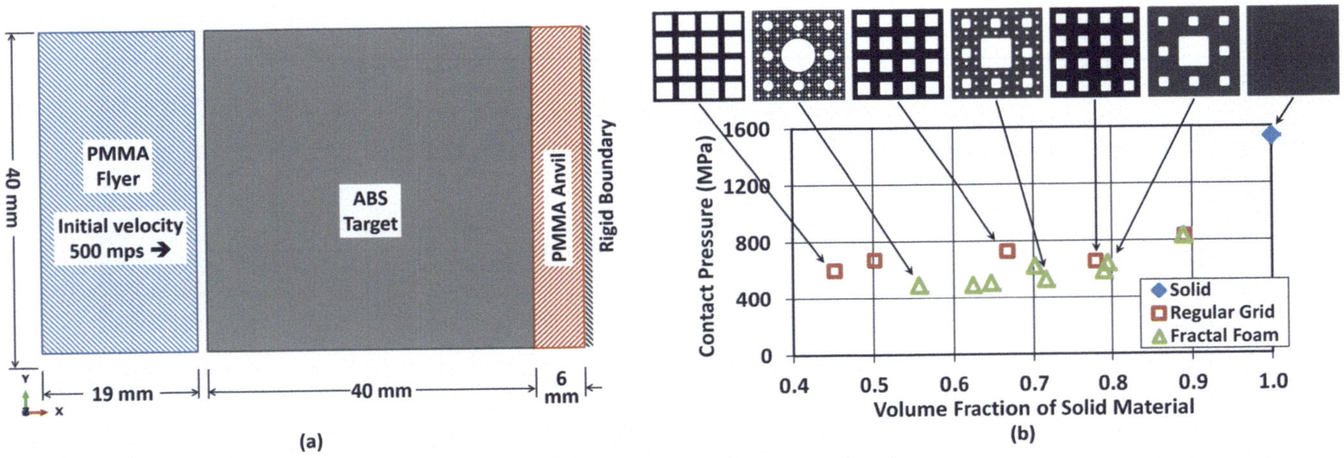

Fig. 10.10 Contact pressure against stiff wall, (**a**) configuration assessed, (**b**) peak contact pressure for fractal foam and regular grid confiurations

A series of targets was assessed to determine contact pressure between the target and anvil. Targets ranged from completely solid ABS material to open grid with less than 50% solid material. Open areas in the targets were modeled as empty voids. Void regions in the targets were either fractal patterns of square or circular regions or were regular grids of square voids.

Time histories of contact pressure were computed by the model, spatially averaged over the entire contact surface. The peak pressure, which generally correlated to the time at which all voids collapsed, were used as the metric for comparing configurations. Results are illustrated in Fig. 10.10b. All configurations with voids had a lower peak contact pressure than a solid target. At any level of solid material volume fraction, the reduction in peak contact pressure for fractal foam target is greater than that for a regular grid target. These results provide some indication that fractal geometries may be more efficient at mitigating shock waves due to impact than other void geometries. Assessment of additional target configurations would need to be performed to ensure that there other void patterns are not more efficient at mitigating shock waves passing through a target. In addition, neither the thickness nor the velocity of the impactor was varied during the modeling runs, rather the impact conditions were fixed. It is possible that these engineered foam structures would be less effective at mitigating longer pressure pulses. These considerations will be included in future studies.

Moreover, the typical nature of the pressure rise on the anvil face suggests a ramp wave propagating in the foam specimens, rather than a shock wave, giving further evidence for shock mitigation in these structures rather than shock propagation. At some distance from the impact face, the incident shock wave is attenuated down to a ramp wave which generates a rising pressure pulse on the fixed anvil, the maximum of which is plotted in Fig. 10.10b. When quantified in this manner, it is interesting to note that at a volume fraction of ~0.9, both the regular grid and fractal foam specimens show the same level of shock attenuation (roughly 50%). It is only at volume fractions ~0.8 and below that the fractal foams exhibit greater shock attenuation than the regular grids, suggesting their suitability as low-density shock absorbing structures.

An additional metric to consider is the time taken for the pressure disturbance to transit the specimen (t_d), compared to the transit time for the shockwave in the solid ABS material (t_s). For a specimen length L, $t_s = L/U_s$, therefore for the 40 mm solid cube specimens, we predict $t_s = 14.0$ μs, using Eq. (10.1) with $u_p = 0.250$ km/s. In Fig. 10.11 below, $U^* = L/t_d$ is plotted vs. volume fraction of ABS for various fractal and regular foam structures. Also plotted on the figure are curves for the p-alpha [21] and Thouvenin [22] models for porous solids. The p-alpha curve was obtained via FEA numerical simulations, using reasonable guesses for the pore compaction behavior. Thouvenin's analysis is a 1-D analysis based on parallel plates with thicknesses and separations which provide the same overall volume fraction as the porous material. In this instance, the agreement with the FEA data is poor. This is most likely due to the inherently 2-D nature of the problem compared with the 1-D analytical model. Agreement with the p-alpha model is much better, especially for low porosities. As the porosity increases towards 50%, however, it is clear that the fractal foams demonstrate much lower disturbance velocities than the regular grid structures.

Experimental data from plate impacts are also plotted in Fig. 10.11. Although only limited experimental data are available for the regular grid geometries, it can be seen that there is very little difference in effective velocity at a volume fraction of 0.89 (geometry shown on right) which corresponds to that of the 1st-order Menger sponge. This is not surprising since the hierarchical geometry of the fractal foams only becomes apparent at higher fractal orders. Also apparent in these data are systematic reductions in experimental velocities from the predicted velocities. This is primarily due to material viscosity and other non-linear effects such as relaxation [23] that are not captured in the FEA model, combined with experimental

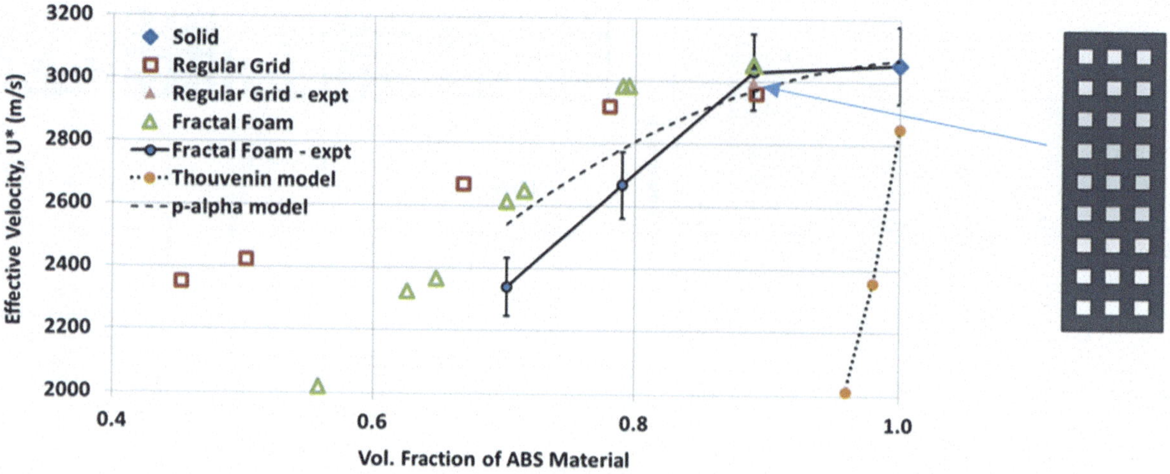

Fig. 10.11 Disturbance velocity, U* plotted vs. volume fraction of material for ABS fractal foam and regular grid configurations; results from FEA modeling and plate impact experiments. Also plotted are predictions from p-alpha [21] and Thouvenin [22] models. Inset on right shows regular grid geometry with same volume fraction as 1st-order fractal foam

challenges for accurately measuring disturbance transit times. For example, the vertical error bars reflect the uncertainty in determining the exact time of impact from the high-speed video data.

The results presented above suggest that there may be reductions in both the speed and the pressure of a disturbance exiting the back face of the fractal foams as compared to a regular (lattice-based) foams at the same volume fraction. Although figures of merit such as these can be subjective, and depend strongly on the loading and boundary conditions, it is possible that fractal foams may offer some advantage for shock mitigation over regular lattice structures. Furthermore, since such a structure might also be asked to be statically load-bearing, the increased second moment of area of the fractal geometry – due to increased mass of material at a distance from the neutral axis – would also be a benefit, especially for structures loaded under in-plane bending.

10.6 Conclusions

The results of the current study suggest that there is an effect of pore geometry on the propagation of compression waves through polymer specimens with engineered porosity. The ability to accurately manufacture 3-D printed objects for use in dynamic impact experiments is clearly an enabler for this kind of study, wherein multiple different geometries can be quickly printed and tested for their impact response. In order to rapidly increase the knowledge base of how engineered foams might respond to shock impact, finite element modeling was used in conjunction with the experimental plate impact testing, at a ratio of 10:1 models to experiments. In order for this approach to be useful, a careful set of experiments was conducted to measure the shock Hugoniot of the solid printed material, allowing calibration of the finite element model. Once established, this finite element model can be used to screen hundreds of potential geometries for their shock response. Furthermore, this approach would lend itself to any one of the available numerical topological optimization schemes, contingent upon the development of suitable and rigorous optimization parameters. From the current study, it is suggested that parameters such as anvil contact pressure or disturbance velocity could be useful, as these both depend on lower-order microstructural details such as volume fraction of solid as well as higher-order features such as hierarchical arrangement of pores *vs.* simple lattice structures. In this regard, it was demonstrated that hierarchical (i.e. fractal) foams behaved more efficiently as shock mitigation structures than did lattice structures, however, the precise origins of their dynamic response have yet to be determined. For example, it has not yet been established whether pore size and spacing have to be of a geometrically fractal nature, or simply polydispersed. Furthermore, it may be that the maximum ligament size, regardless of whether the geometry is fractal, governs the chosen figures of merit, and the influence of loading duration on disturbance transit time has yet to be established. Further experiments are planned in this direction, including the use of PCI under impact conditions to determine whether directionality of shock response can be engineered into these kinds of structures, i.e. the generation of "shock diode" characteristics where the propagation of true shocks can be accomplished in one direction but not in the opposite direction.

Acknowledgements This work was supported by Air Force Office of Scientific Research, 6.1 LRIR "Novel Material Microstructures for Shock Survivability (16RXCOR326)", Dynamic Materials and Interactions Portfolio, Program Officers Dr. Jennifer L. Jordan, Dr. Martin J. Schmidt, AFOSR/RT. Experiments at the Dynamic Compression Sector were performed by Los Alamos National Laboratory at Argonne National Laboratory's Advanced Photon Source (APS). The authors acknowledge support from the Laboratory Directed Research and Development (LDRD) program at Los Alamos National Laboratory (Project No. 20160103DR) and DOE/NNSA Campaign 2. LANL is operated by Los Alamos National Security, LLC, for the U.S. Department of Energy (DOE) under Contract No. DEAC52- 06NA25396. Use of the Advanced Photon Source, an Office of Science User Facility operated for the DOE Office of Science by Argonne National Laboratory, was supported by the U.S. DOE under Contract No. DE-AC02-06CH11357. Cleared for Public Release, Distribution Unlimited 88ABW-2018-0733.

References

1. https://en.wikipedia.org/wiki/Mudcrack. Accessed 2/8/2018
2. http://www.monkeybuddha.blogspot.com/2017/05/fractal-wood-burning.html. Accessed 2/8/2018
3. http://depositphotos.com/search/bullet-glass-st200.html?qview=15344565. Accessed 2/8/2018
4. Mandelbrot, B.B.: The Fractal Geometry of Nature. Macmillan (1983)
5. Menger, K.: Allgemeine Räume und Cartesische Räume. I., Communications to the Amsterdam Academy of Sciences. English translation reprinted in Edgar, Gerald A., ed. (2004), Classics on fractals, Studies in Nonlinearity, Westview Press. Advanced Book Program, Boulder, CO, ISBN 978-0-8133-4153-8, MR 2049443 (1926)
6. http://www.wolframalpha.com/input/?i=Menger+sponge. Accessed 2/8/2018
7. Gibson, L.J., Ashby, M.F.: Cellular Solids: Structure and Properties, 2nd edn. Cambridge University Press, Cambridge
8. Sheffield S., Gustavsen R., Alcon R.: In-Situ Magnetic Gauging Technique Used at LANL- Method and Shock Information Obtained. In: Shock Compression of Condensed Matter – 1999, Snowbird, Utah (2000)
9. Gustavsen R., Sheffield S., Alcon R., Hill L.: Shock Initiation of New and Aged PBX 9501 Measured with embedded Electromagnetic Particle Velocity Gauges, LA-13634-MS, Los Alamos National Laboratory
10. Branch, Brittany (private communication, April 2017)
11. Jensen, B.J., et al.: Ultrafast, high resolution, phase contrast imaging of impact response with synchrotron radiation. AIP Adv. **2**(1), 6 (2012)
12. Jensen, B.J., et al.: Impact system for ultrafast synchrotron experiments. Rev. Sci. Instrum. **84**(1), 5 (2013)
13. Jensen, B.J., et al.: Dynamic experiment using IMPULSE at the Advanced Photon Source. In 18th APS-SCCM and 24th AIRAPT, Pts 1–19, W. Buttler, M. Furlanetto, and W. Evans, Editors. 2014, IOP Publishing Ltd: Bristol
14. Abaqus Analysis User's Guide, Version 6.14, Dassault Systèmes (2014)
15. Lacina, D., Neel, C.: Determination of RX-supplied, solid (no intentional voids) AM material. UDRI/AFRL draft report, December 2016
16. Nakai, K., Yokoyama, T.: High strain-rate compressive properties and constitutive modeling of selected polymers. J. Solid Mech. Mater. Eng. **6**(6), 731–741 (2012)
17. Peterson, A., et al.: Dynamic Evaluation of Acrylonitrile Butadiene Styrene Subjected to High-Strain-Rate Compressive Loads ARL-TN-0648, Army Research Laboratory, Dec. 2014
18. Steinberg, D.J.: Equation of State and Strength Properties of Selected Materials. UCRL-MA-106439 (Change 1), Lawrence Livermore National Laboratory (1996)
19. Frank, G.: Analytic and Experimental Evaluation of the Effects of Temperature and Strain Rate on the Mechanical Response of Polymers. UDR-TR-97-152, University of Dayton Research Institute, December, 1997
20. Branch, B., Ionita, A., Clements, B., Montgomery, D.M., Jensen, B.J., Patterson, B.M., Schmalzer, A., Mueller, A., Dattelbaum, D.M.: Controlling shockwave dynamics using architecture in periodic porous materials. J. Appl. Phys. (2017). https://doi.org/10.1063/1.4978910
21. Herrmann, W.: Constitutive equation for the dynamic compaction of ductile porous materials. J. Appl. Phys. **40**, 2490–2499 (1969)
22. Thouvenin, J.: Effect of a Shock Wave on a Porous Solid, in Proc. Fourth Symposium on Detonation, Naval Ordinance Laboratory, Silver Spring, MD, 1965, p. 258
23. Schuler, K.W., Nunziato, J.W., Walsh, E.K.: Recent results in nonlinear viscoelastic wave propagation. Int. J. Solids. Struct. **9**, 1237–1281 (1973)

Chapter 11
Mechanical and Thermal Characterization of Fused Filament Fabrication Polyvinylidene Fluoride (PVDF) Printed Composites

Niknam Momenzadeh, Carson M. Stewart, and Thomas Berfield

Abstract Polyvinylidene fluoride (PVDF) is a polymer that offers a variety of desirable material properties. Its high resistance to corrosive acids and its capability to show piezoelectric behavior are some of these properties that are attractive to many industrial applications. Three-dimensional printing of PVDF is extremely difficult using fused filament fabrication processes due to the large coefficient of thermal expansion of homopolymer PVDF, which results in substantial component warping. In the present work, the effect of zirconium tungstate microparticles as a secondary phase within a PVDF matrix is experimentally studied. Viable printing parameters and the corresponding mechanical and thermal behavior of the PVDF composite structures based on digital image correlation tests are presented.

Keywords Additive manufacturing · Fused filament fabrication · Coefficient of thermal expansion · Digital image correlation · Tensile test

11.1 Introduction

Additive Manufacturing (AM) is a technology in which a digital three-dimensional (3D) model is utilized to create a part geometry by joining the constituent materials layer-upon-layer. According to ASTM F2792-12a [1], AM is divided into seven different categories. As one of the AM categories, material extrusion is a process in which components are fabricated by extruding molten thermoplastic material, high viscosity slurries, or low melting temperature metals through a nozzle or an orifice that moves in x, y, z directions. Fused Filament Fabrication (FFF) is the first and the most popular extrusion-based method that can extrude distinct materials using multiple extrusion nozzles. FFF is portable, easily scalable, and a cost effective method which can print multiple materials through a straightforward mechanism [2].

Piezoelectricity is a material characteristic that is described as the ability of producing electrical charge when subjected to mechanical stress or vibration (or vice versa). This property can be found in natural materials such as quartz, Rochelle salt, topaz, bone, dentin, and hair or man-made materials like barium titanate or lead titanate. In the late 1800's, Jacques and Pierre Curie found that quartz crystals would generate electric charges under applied pressure. The converse effect of piezoelectric phenomenon (stress generation by applying electrical field) was investigated by Lippman in 1881 [3]. Polymers that demonstrate piezoelectricity generally demonstrate lower electro-mechanical coupling coefficients than their ceramic counterparts, but the greater strain tolerance of piezoelectric polymers make them highly useful in certain applications. Some of the popular applications for piezoelectric materials include energy harvesting systems, actuators, sensors [4], transducers [5], piezoelectric motors, medial applications [4, 6], and driving vibrations [7]. PVDF is a semi-crystalline thermoplastic polymer that has a strong piezoelectric response for the β-phase microstructure orientation. Piezoelectric properties of properly oriented PVDF was discovered by Kawai in 1969 [8]. Due to the low cost, low melting point and potential for piezoelectric properties [9], PVDF is an intriguing choice for use in FFF machines.

Warping deformation due to residual stresses and thermal gradients is one of the common challenges in fabrication of parts via FFF process. The warping deformations are often caused by uneven heat distribution within a part during the printing process and typically lead to significant dimensional inaccuracies in the fabricated components. The thermal stresses that develop during melt solidification can prompt warping [10], but these effects are exacerbated in the presence of non-optimal printing parameters, certain geometrical (aspect-ratio) designs, or specific material properties [11, 12] like larger coefficients of thermal expansion. Although some research has been conducted on the theoretical modeling of warping deformations

N. Momenzadeh (✉) · C. M. Stewart · T. Berfield
University of Louisville, Louisville, KY, USA
e-mail: niknam.momenzadeh@louisville.edu

S. Kramer et al. (eds.), *Mechanics of Additive and Advanced Manufacturing, Volume 8*, Conference Proceedings of the Society for Experimental Mechanics Series, https://doi.org/10.1007/978-3-319-95083-9_11

during FFF, there are very limited methodical studies in this field [10]. In this research, a method is presented for controlling part warping during the PVDF printing process. In this method, zirconium tungstate ($Zr(WO_4)_2$) microparticles with a negative coefficient of thermal expansion are added to a PVDF matrix as a secondary phase prior to filament extrusion. This study concentrates on the effects of particulate-loading concentration on component warping. Additionally, The mechanical and thermal characteristics of the PVDF and PVDF/zirconium tungstate specimens fabricated via FFF process are experimentally probed using digital image correlation system and standard tensile testing.

11.2 Experimental Methods

11.2.1 Materials

Resin pellets of homopolymer PVDF Kynar 740 from Trident Engineering Plastics Inc. were utilized as the basic material. A custom filament extrusion device was used in order to extrude the pellets and produce 1.75 ± 0.15 mm diameter filament of PVDF. The filament extrusion head temperature for filament fabrication was set at 215 °C. Zirconium Tungstate powder from American Elements was used as the secondary constituent for creating PVDF composite filaments. Zirconium tungstate microparticles with negative coefficient of thermal expansion (CTE) was selected for mitigating the total CTE of PVDF composite structures, with the intent to reduce warping issues during FFF printing. One batch of PVDF composite with 2.5% Zirconium tungstate by volume was prepared.

11.2.2 Equipment

A Reprap Prusa i3 3D printer was used for fused filament fabrication of homopolymer PVDF and PVDF composite parts. Filaments of materials were extruded through a Bowden extruder. All the samples were printed on a prepared bedplate that was initially cleaned using acetone and covered by 3M blue painters tape. Materials were deposited through a sharp nozzle with 0.35 mm diameter.

A torque rheometer machine, Intelli-Torque Plasti-Corder®, was utilized for mixing PVDF pellets and zirconium tungstate powder. The speed was set at 10 rpm for making a mixture of homopolymer PVDF and zirconium tungstate. One batch of PVDF composite with 2.5% zirconium tungstate by volume was produced by torque rheometer machine. Afterwards, by taking advantage of a chopper machine, some pellet shape of particular PVDF composite were generated which was utilized in creating PVDF composite filament via the filament extruder.

11.2.3 Processing

Samples for tensile testing were designed according to ASTM D1708 as shown in Fig. 11.1. For tensile and thermal expansion experiments, Five and two test specimens of homopolymer PVDF and PVDF composite were examined, respectively. The thickness of samples was 1.5 mm, which corresponded to ten deposited layers for each test specimen.

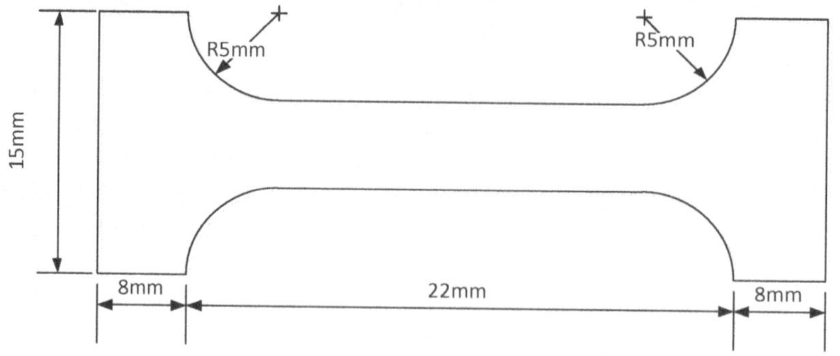

Fig. 11.1 Schematic representation of tensile test specimens

Table 11.1 Printing parameters

Printing parameters	Value	Printing parameters	Value
Layer height	0.15 mm	Speed (perimeters)	10 mm/s
First layer height	0.2 mm	Speed (infill)	15 mm/s
Perimeters	2	Speed (top solid)	10 mm/s
Fill density	70%	Nozzle diameter	0.35 mm
Fill pattern	Rectilinear	Nozzle temperature	210 °C
Brim	3	Bed temperature	90 °C

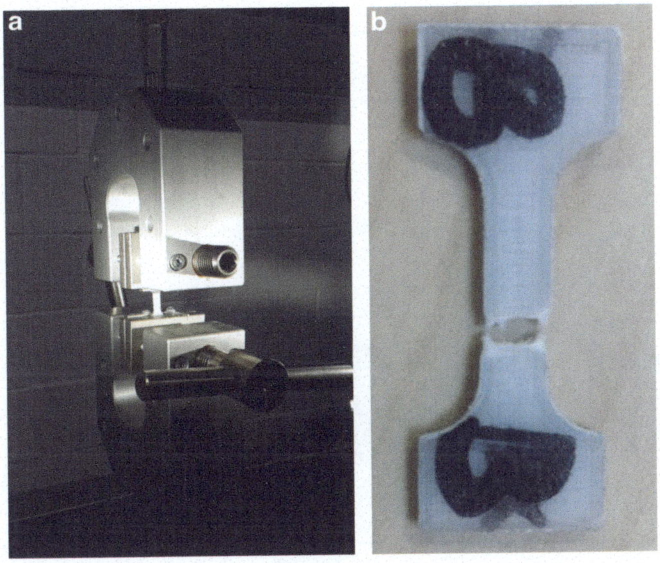

Fig. 11.2 Tensile testing machine grips (**a**), and typical sample after testing (**b**)

3D printing of PVDF and PVDF composite was the challenging part of this study. Gaps between infill and perimeters, poor adhesion between printed layers, stringy infill, and clogged nozzle were some of the problems that were observed during printing. By modifying the parameters and evaluating the outcomes on the printed parts, optimal parameters were obtained for successful printing. The influences of deposition parameters on the quality and integrity of components and their mechanical and thermal properties in different AM methods were studied by many recherches [13–15]. Test specimens were printed under the viable printing parameters which are shown in Table 11.1.

11.3 Result and Discussion

11.3.1 Tensile Test

Each additively manufactured FFF sample was tested in tension in a MTI universal testing machine. The crosshead velocity was set at 0.05 inch/min and the upper grip position was incremented until fracture at room temperature. The tensile test apparatus and a representative sample fractured after testing are shown in Fig. 11.2.

Maximum load was recorded during testing for both homopolymer PVDF and PVDF composite as shown in Figs. 11.3 and 11.4. The maximum load in homopolymer PVDF (pure PVDF) samples was 298.32 N and for PVDF composite was 195.5 N. The results for all five tests were averaged to determine the maximum load as shown in the tables below, while the average maximum load for pure PVDF is 276.81 N, this value for PVDF composite equals to 192.2 N. Hence, PVDF composites using zirconium tungstate as a filler were found to have significantly lower mechanical strength than pure PVDF.

Figure 11.5 shows the load-displacement plot for all the samples. From the figure, the total energy absorption prior to fracture in pure PVDF is much higher than that in the PVDF composites.

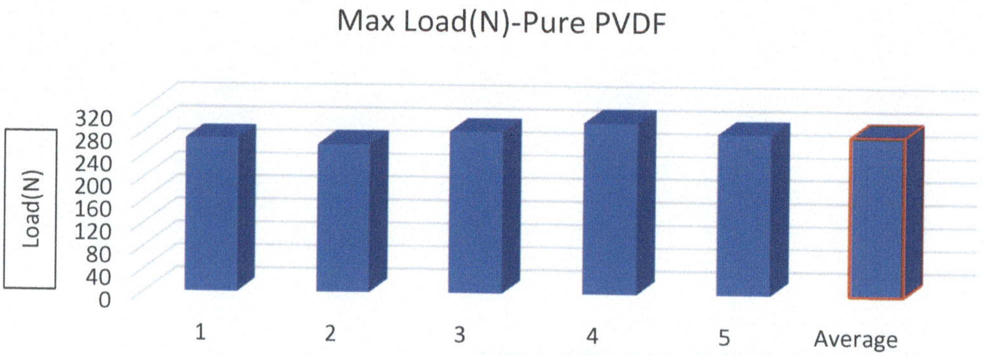

Sample	1	2	3	4	5	Average
Max Load(N)	267.5	257.51	281.04	298.34	280.01	276.81

Fig. 11.3 Maximum load for homopolymer PVDF

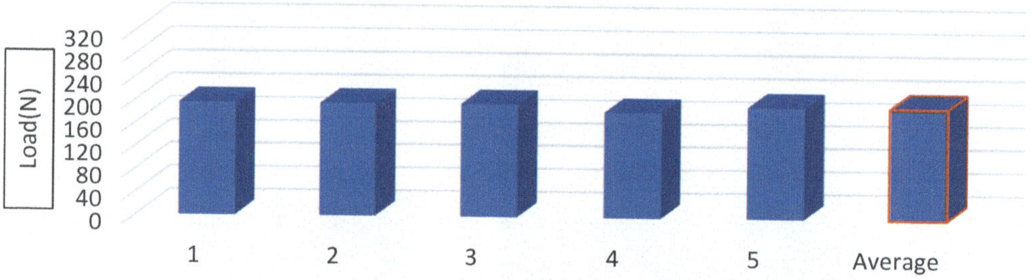

Sample	1	2	3	4	5	Average
Max Load(N)	195.45	194.82	195.08	182.77	192.93	192.21

Fig. 11.4 Maximum load for PVDF composite

11.3.2 Thermal Expansion Testing

Coefficient of thermal expansion (CTE) is a material characteristic that describes the length variation of that material when the temperature is increased or decreased. If α = coefficient of thermal expansion (assuming linear behavior over the probed temperature range), ΔT = Temperature differential and $\Delta L/L$ = thermal strain, then the thermal expansion coefficient is as follows:

$$\alpha = (1/\Delta T)\,(\Delta L/\mathrm{L}) \tag{11.1}$$

Hence, by knowing the amount of temperature differences and thermal strain, CTE can be calculated. Although, the majority of materials exhibit positive thermal expansion upon heating, a group of material contracts in the same temperature change. Due to this unique feature, this group of material has essential applications in composite materials. One of this kind of material is zirconium tungstate, with CTE around -9×10^{-6}/K for temperatures ranging from 0.3 to 1443 K [16].

The coefficients of thermal expansion of both pure PVDF and PVDF composite test specimens were calculated based on Eq. (11.1). Thermal strain in this study was measured using a digital image correlation (DIC) system. DIC is an optical non-contact method for measuring deformation and strain. The images were taken by two Nikon D3200 Cameras. Two speckled pattern images are shown in Fig. 11.6.

A DATAPLATE Hot Plate/Stirrer Series 720 was used as the thermal source for the CTE experiments. The temperature of the plate was increased from room temperature to 180 °C. Samples were placed on the surface of the hot plate and a FLUKE 179 true RMS multimeter was used to measure the surface temperature of the part. Images of the surface were acquired by

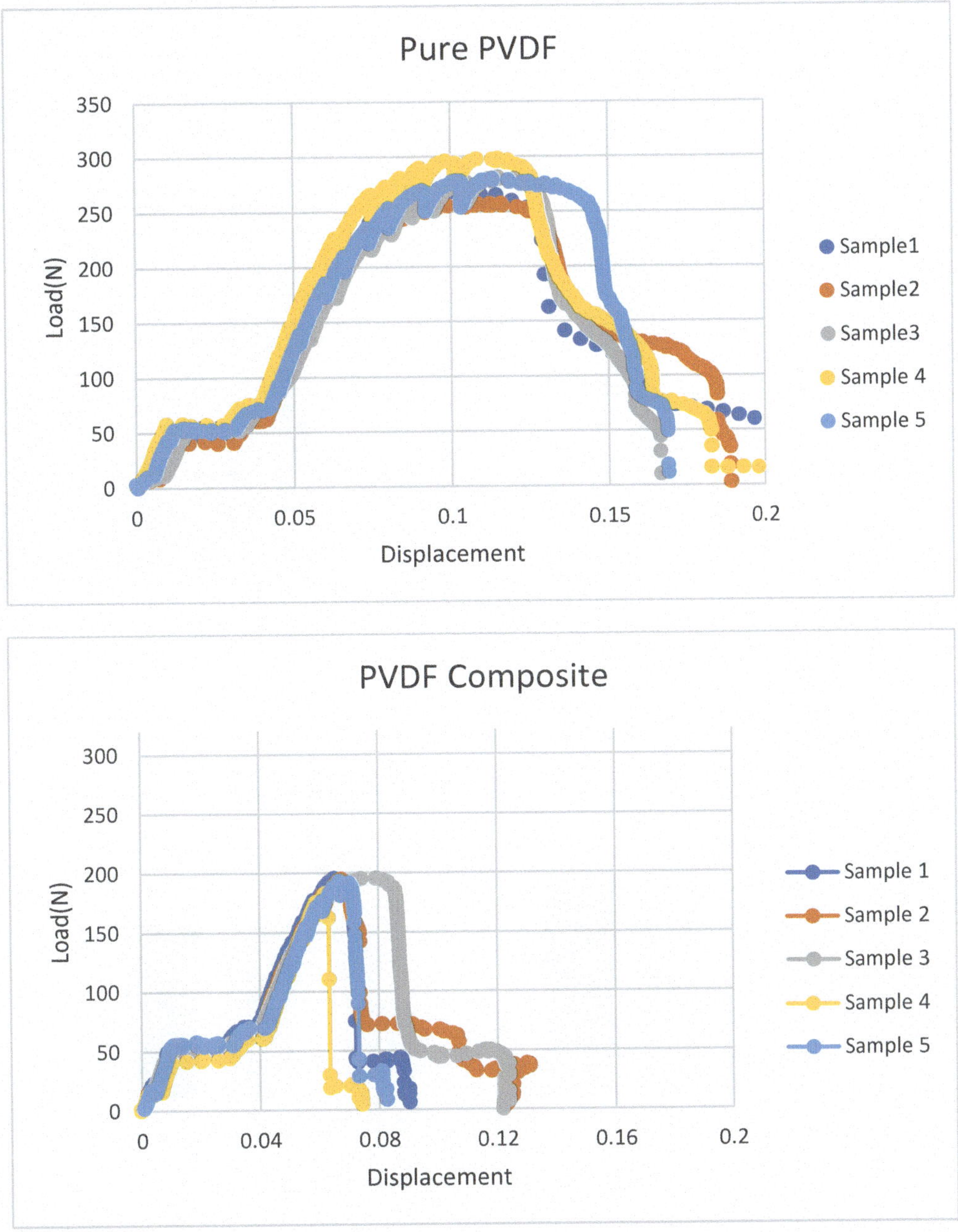

Fig. 11.5 Load versus displacement plot of pure PVDF (top) and PVDF composite (bottom) specimens

high-resolution digital cameras at a rate of 1 image per 10s. VIC-3D (Correlated Solutions) software was used to analyze the captured images and to calculate the thermal strain state. The maximum thermal strain experienced for the specimens during testing are shown in Fig. 11.7. The maximum thermal strain in PVDF composite is lower than pure PVDF.

In addition, using the Eq. (11.1) and appropriate amount of thermal strain and ΔT obtained from the experiments, the coefficient of thermal expansion of the samples were determined as shown in Fig. 11.8.

The PVDF composite specimens had a lower CTE than pure PVDF, which means thermal deformations in the composite are much smaller than homopolymer over the examined temperature range. Hence, it can be concluded that, adding zirconium tungstate to homopolymer PVDF in FFF printing, will decrease the probability of deformations due to thermal warping and residual stresses.

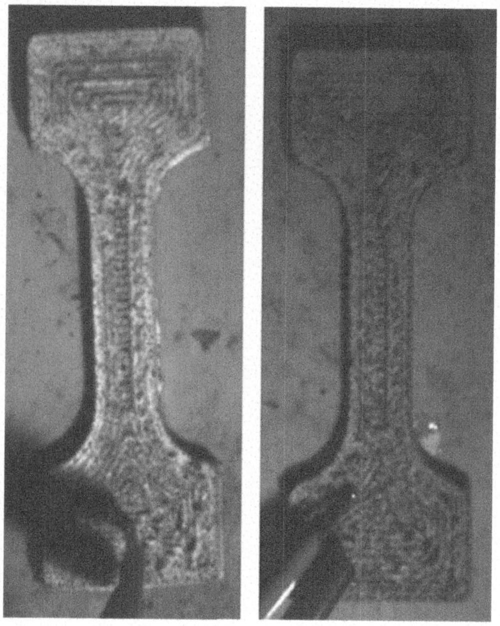

Fig. 11.6 DIC image showing speckled homopolymer PVDF (left), and PVDF composite (right) specimens used for thermal performance testing

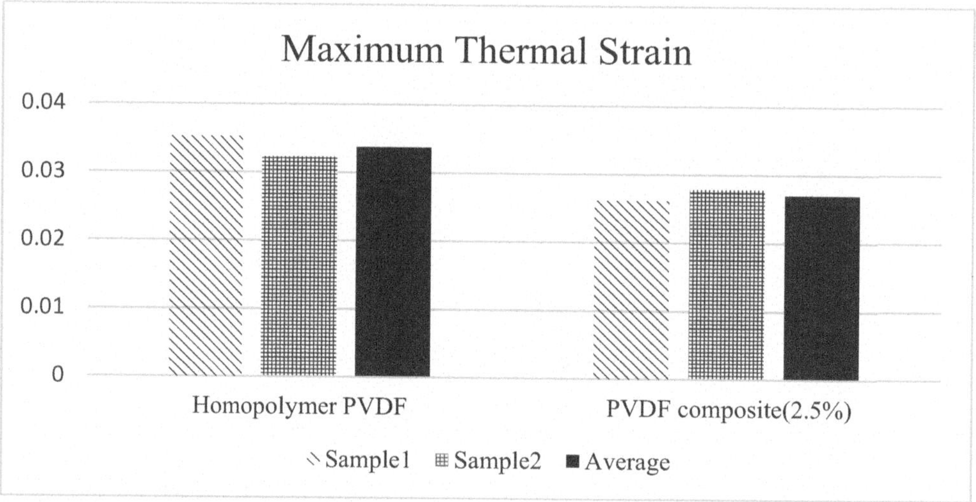

Fig. 11.7 Maximum thermal strain in pure PVDF and 2.5% $(Zr(WO_4)_2)$/PVDF composite

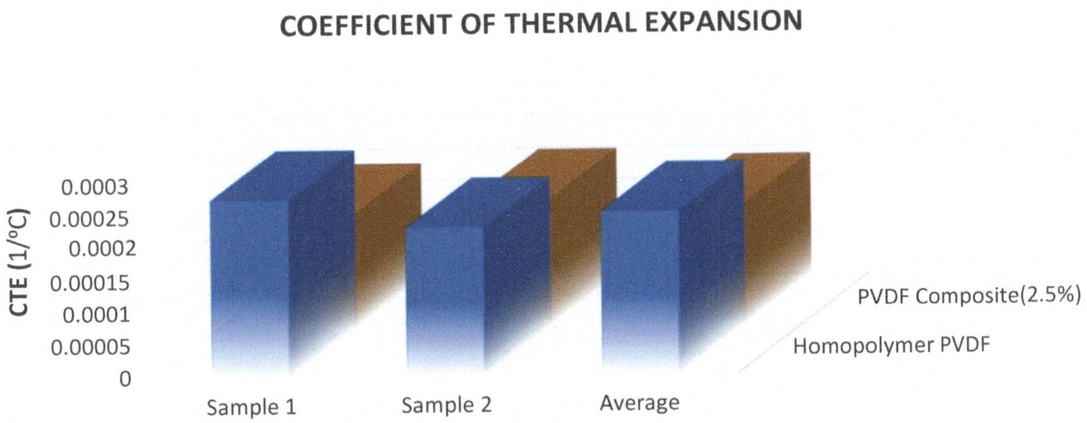

Fig. 11.8 Coefficient of thermal expansion of pure PVDF and PVDF composite

11.4 Conclusion

In the present study, the effect of zirconium tungstate on different characteristics of the components fabricated via FFF process was investigated in the attempt to decrease the commonly encountered issue of part warping due to thermal gradients and residual stresses. It was concluded that, although the homopolymer PVDF demonstrates a much higher failure stress than the $(Zr(WO_4)_2)$/PVDF composite, the negative coefficient of thermal expansion of zirconium tungstate helps to lower the effective composite CTE by approximately 20%. This influence makes zirconium tungstate a good candidate as a filler for part fabrication using FFF 3D printing of PVDF.

References

1. American Society for Testing and Materials (ASTM), Standard Terminology for Additive Manufacturing Technologies(F2792–12). http://www.astm.org, 2015
2. Gibson, I., Rosen, D.W., Stucker, B.: Additive Manufacturing Technologies: Rapid Prototyping to Direct Digital Manufacturing, p. 462. Springer Publishing Company, Inc, Boston (2009)
3. Dineva, P., et al.: Piezoelectric materials. In: Dynamic Fracture of Piezoelectric Materials: Solution of Time-Harmonic Problems via BIEM, pp. 7–32. Springer International Publishing, Cham (2014)
4. Bar-Cohen, Y., Zhang, Q.: Electroactive polymer actuators and sensors. MRS Bull. **33**(3), 173–181 (2008)
5. Khaled, S.R., Sameoto, D., Evoy, S.: A review of piezoelectric polymers as functional materials for electromechanical transducers. Smart Mater. Struct. **23**(3), 033001 (2014)
6. Naohiro, M., Hiroshi, O.: Piezoelectric polymers and their applications. Jpn. J. Appl. Phys. **22**(S3), 3 (1983)
7. Derakhshani, M., Berfield, T., Murphy, K.D.: Dynamic Analysis of a Bi-stable Buckled Structure for Vibration Energy Harvester. Springer International Publishing, Cham (2018)
8. Heiji, K.: The piezoelectricity of poly (vinylidene fluoride). Jpn. J. Appl. Phys. **8**(7), 975 (1969)
9. Porter, D.A., Hoang, T.V.T., Berfield, T.A.: Effects of in-situ poling and process parameters on fused filament fabrication printed PVDF sheet mechanical and electrical properties. Addit. Manuf. **13**, 81–92 (2017)
10. Turner, B.N., Gold, S.A.: A review of melt extrusion additive manufacturing processes: II. Materials, dimensional accuracy, and surface roughness. Rapid Prototyp. J. **21**(3), 250–261 (2015)
11. Wang, T.-M., Xi, J.-T., Jin, Y.: A model research for prototype warp deformation in the FDM process. Int. J. Adv. Manuf. Technol. **33**(11), 1087–1096 (2007)
12. Herrmann, K.-H., et al.: 3D printing of MRI compatible components: Why every MRI research group should have a low-budget 3D printer. Med. Eng. Phys. **36**, 1373–1380 (2014)
13. Lanzotti, A., Grasso, M., Staiano, G., Martorelli, M.: The impact of process parameters on mechanical properties of parts fabricated in PLA with an open-source 3-D printer. Rapid Prototyp. J. **21**(5), 604–617 (2015)
14. Miyanaj, H., Momenzadeh, N., Yang, L.: Effect of printing speed on quality of printed parts in Binder Jetting Process. Additive Manufacturing (2017)
15. Afrose, M.F., Masood, S.H., Iovenitti, P., Nikzad, M., Sbarski, I.: Effects of part build orientations on fatigue behaviour of FDM-processed PLA material. Prog. Add. Manufact. **1**(1–2), 21–28 (2016)
16. Sullivan, L.M., Lukehart, C.M.: Zirconium tungstate (ZrW2O8)/polyimide nanocomposites exhibiting reduced coefficient of thermal expansion. Chem. Mater. **17**(8), 2136–2141 (2005)

Chapter 12
Influence of an Extreme Environment on the Tensile Mechanical Properties of a 3D Printed Thermoplastic Polymer

Jose Torres, Otito Onwuzurike, Amber J. W. McClung, and Juan D. Ocampo

Abstract Inspired by the concept of deploying 3D printers into the field to produce parts on demand, the purpose of this study is to examine the effects of the environment on a 3D printing apparatus and its ability to produce consistent operative prints. The inelastic deformation behavior of Poly-lactic Acid, a biodegradable thermoplastic, 3D printed in extreme environments was investigated. The experimental program was specifically designed to explore the influence of ambient temperature (25–40 °C) during the printing process on the mechanical performance of the printed material. In order to understand the effects of the printing environment versus general exposure to extreme environments, samples were also printed at 25 °C and subsequently aged in an oven at temperatures ranging from 30 to 40 °C before mechanical testing. All mechanical testing was performed in standard laboratory temperature and humidity. The influence of the print temperature and oven aging on the elastic modulus, yield stress, strain energy, and tensile stress are all compared. In addition, the capacity to accumulate strain before failure is compared.

Keywords 3D printing · Environmental aging · Print variations · Poly-lactic acid · Tensile testing

12.1 Introduction

Digitally manufacturing methods, such as 3-D printing, are promoted as revolutionizing engineering design and repair. Using advanced imaging techniques and computer-aided design, efforts are underway in the engineering and technology community to create personalized components ranging from prototypes, repair parts, prosthetics, biological tissue, and even food. This method of production is beneficial because parts need to be readily available on site. Logistically speaking, it is complex and costly to ship a part that is needed for a field repair across the globe in a timely manner. Also, having parts on hand takes up space in transportation. Space that can be utilized more efficiently through printing on site. Hence, the United States military has moved toward placing 3-D printers with deployed troops (including in extreme environments) [1, 2].

To assure long-term durability and structural integrity of the digitally manufactured components created in these extreme environments, a thorough understanding of the environmental effects on a 3-D printing apparatus is essential. Of particular interest is a hostile environment where temperatures are high, as these parts will be printed and utilized in a similar setting. The hot environment causes the filament to be heated prior to being formed into the object which may result in material properties unlike those specified by the manufacturer.

The research objective of this study is to investigate the influence extreme temperatures on the mechanical behavior of 3d-printed polymers. The effects of extreme temperatures during printing is investigated as well as the effect of thermal aging after the printing process is complete. Experimental results generated here will give insight into the changes in the inelastic behavior of the polymer with prior aging time and provide a foundation for extending constitutive equations to capture the effects of prior aging on the deformation behavior. These materials are promoted as revolutionizing engineering design but comprehensive characterization plans are yet to be developed to understand the array of material properties that can be produced during printing. To assure long-term durability and structural integrity of the digitally manufactured components, a thorough understanding of the effects of the exposure to degrading environment on the mechanical behavior of the printed material is essential.

J. Torres · O. Onwuzurike · A. J. W. McClung (✉) · J. D. Ocampo
Department of Engineering, St. Mary's University, San Antonio, TX, USA
e-mail: amcclung@stmarytx.edu

S. Kramer et al. (eds.), *Mechanics of Additive and Advanced Manufacturing, Volume 8*, Conference Proceedings of the Society for Experimental Mechanics Series, https://doi.org/10.1007/978-3-319-95083-9_12

12.2 Methodology

This research project aims to understand the environmental effects on the performance 3-D printed polymers. To obtain reliable comparative results, we used polymer filament printers of the same make and model, ANET A8 Desktop ED Printers from Prusa i3. The print settings included a layer height of 0.2 mm, shell thickness of 1.6 mm and a fill density 100. A pair of identical printers were used in this project. Two AcuRite 00613 Indoor Humidity Monitors where used to monitor the temperature and humidity during the time of printing.

The material printed is a 1.75 mm diameter PLA (poly-lactic acid) Filament manufactured by K-Camel. The material was printed into a dogbone shape following ASTM D638 Type IV (total length = 115 mm, overall width = 19 mm, gage length = 25 mm, width of narrow section = 6 mm). The baseline dogbones were printed in standard laboratory atmosphere (25 °C). The second set of dogbones was printed in an environment with an elevated temperature in sets at 30, 35, and 40 °C. The third set of dogbones was also printed in a standard laboratory environment (25 °C) and subsequently aged at 30, 35, and 40 °C for 5 h and 15 min to match the amount of time that the printing process lasted for the second batch of dogbones.

All three sets of polymers were tested in uniaxial tension on an MTS Criterion Model 43 with a 30kN load cell. All tests were conducted in crosshead control mode with a rate of 0.01 mm/sec. Strains were measured using an MTS Model 634.11 Axial Extensometer. Reported strains and stresses following the engineering definitions, strain = change in gage length/initial gage length and stress = measured force/initial cross sectional area.

12.3 Experimental Observations

Specimens were subjected to displacement-controlled tension-to-failure tests at 25 °C at a constant displacement rates of 0.01 mm/sec. At least three samples were tested at each rate for each dogbone set. The axial stress–strain responses of a representative specimen from the baseline group (printed at room temperature and not aged), group 2 (printed at 40 °C and not aged), and group 3 (printed at room temperature and aged at 40 °C) are all shown in Fig. 12.1.

For the baseline behavior (listed as Room temperature in the figure), the axial stress–strain curves exhibit an overall average quasi-elastic slope (E) upon leaving the origin. The initial elastic modulus (or Young's modulus) is observed to decrease with oven aging and with printing at elevated temperature. The maximum stress values are observed to increase with both oven aging and with printing at elevated temperature. And the strain at fracture is observed to decrease with both oven aging and with printing at elevated temperature. Current work will result in graphically compiling the results of printing at each of the elevated temperatures in one figure. In addition, the analysis will compile the results of aging at each temperature all in one figure.

Ongoing analysis is quantifying the observed differences in the monotonic tensile behavior in the three sets of conditions outlined above. The elastic slope will be measured from 0 to 1.5% strain. The yield stress will be measured using a 0.002 offset. The maximum stress, the strain at fracture, and the stress at fracture will also be quantified and compared. All of this analysis will analyze the trends comparing the baseline behavior to the samples printed at 30, 35, and 40 °C. The same analysis

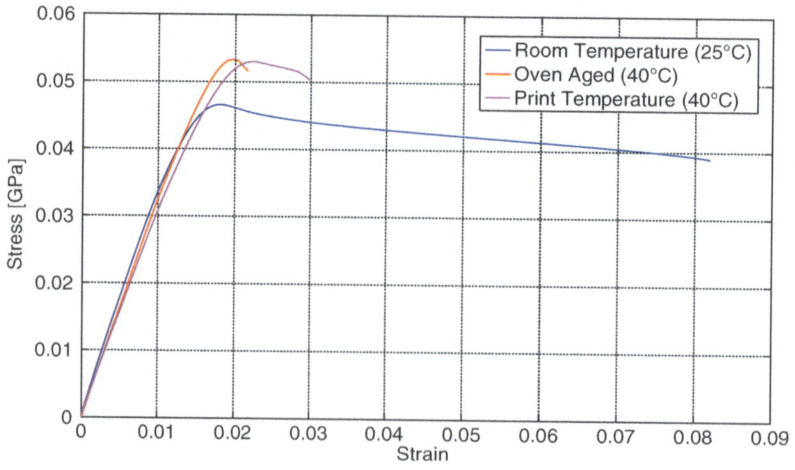

Fig. 12.1 Monotonic tension to failure for a baseline sample versus a sample printed at 40 °C and a sample aged at 40 °C

will compare the trends for samples printed at 25 °C and then aged for 5 h and 15 min at 30, 35, and 40 °C. The results will be compared to elucidate which properties change due to the material being stored in an elevated temperature environment and which properties only change when the actual printing is conducted in the elevated temperature environment.

Acknowledgements This research was conducted with the support of the office of Sponsored Projects, Academic Research, & Compliance St. Mary's University.

References

1. Department of Defense, 3Dprint.com: the voice of 3D printing/additive manufacturing [Online]. Available: https://3dprint.com/tag/department-of-defense/. Accessed 26 Feb 2018
2. Marines take 3D printed drones from teh lab to the field, Defense Systems [Online]. Available: https://defensesystems.com/articles/2017/05/08/marinecorpprint.aspx. Accessed 26 Feb 2018

Chapter 13
A Framework for Estimating Mold Performance Using Experimental and Numerical Analysis of Injection Mold Tooling Prototypes

Suchana Jahan, Hazim El-Mounayri, Andres Tovar, and Yung C. Shin

Abstract Additive Manufacturing (AM), 3D printing, rapid prototyping, or rapid tooling refer to a range of technologies that are capable of translating virtual CAD model data into physical model. It is executed in growing number of applications nowadays. A wide range of materials are currently being used to produce consumer products and production tools. AM has brought in revolutionary changes in traditional manufacturing practices. Yet, there are certain drawbacks that hinder its advancement at mass manufacturing. High cost associated with AM is one of them. Using 3D printed tooling can provide long-time cost effectiveness and better product quality. Additively manufactured injection molds can increase the cooling performance, reduce production cycle time, and improve surface finish and part quality of the final plastic product. Yet, manufacturers are still not using the printed molds for industrial mass production. Numerical analysis can provide approximation of such improved performance, but, factual experimental results are necessary to satisfy performance criteria of molds to justify the large investment into tooling for existing industries. In this research work, a desktop injection molding machine is used to evaluate performance of 3D printed molds to develop a cost and performance analysis tool. It serves as a baseline to predict the performance of molds in real-time mass manufacturing of consumer products. The analysis describes how appropriate the estimation can be from any simulation study of molds, how much the scaling down of tool and molding system can affect the prediction of actual performance, what correction factors can be used for better approximation of performance matrices. Several "scaled down" prototypes of injection molds have been used. They have design variations as: with or without cooling system, conformal or straight cooling channels, solid or lattice matrix, and metal or tough resin as the mold material. The molds are printed in in-house printing machines and can also be printed online with limited charges. This also provides an excellent demonstration of using inexpensive material and manufacturing process, such as resin to estimate the performance of highly expensive 3D printed stainless steel molds. The work encompasses a framework to reduce overall cost of implementing AM, by lowering time and monetary expenses during the research and development, and prototyping phases.

Keywords 3D printing · Injection molding · Scaled down prototype · Performance comparison · Conformal cooling

13.1 Introduction

In recent years, additive manufacturing has become the buzzword in manufacturing world. People are trying to build anything and everything with additive manufacturing, commonly known as 3D printing. With its versatile potential to build a lot of critical and complex geometry, this technology is bringing in revolutionary changes in the manufacturing industry. But cost of 3D printing becomes a major hindrance for this technology to be used in mass production.

One of the prospective fields of application of 3D printing is in the plastic injection molding industry. The current practice of this industry is to use traditional manufacturing operations, such as milling, drilling, grinding etc. to make the molding tools and use them for the final production of plastic parts in the plant. 3D printing can take over the production of tools and provide optimally designed highly effective molds that can offer long-term business benefits. But this is a difficult decision for the mold makers to move out from their traditional practice and invest into the additive manufacturing of the molding tools. Major difficulty lies in predicting and testing the mold before going into mass scale production. This situation led into trying out a

S. Jahan (✉) · Y. C. Shin
School of Mechanical Engineering, Purdue University, West Lafayette, IN, USA
e-mail: sjahan@purdue.edu

H. El-Mounayri · A. Tovar
School of Engineering and Technology, IUPUI, Indianapolis, IN, USA

S. Kramer et al. (eds.), *Mechanics of Additive and Advanced Manufacturing, Volume 8*, Conference Proceedings of the Society for Experimental Mechanics Series, https://doi.org/10.1007/978-3-319-95083-9_13

low cost way to predict the mold behavior. This paper presents a solution to this situation and establishes a framework to lead the pathway towards a cost -effective analysis and prediction tool to help taking fast decisions in molding industry.

13.2 Methodology

This paper presents a basic framework towards creating an accurate performance prediction tool of plastic injection molds and overall plastic injection molding technique. With the advances in engineering and technology in present age, it is a common practice to predict the behavior of any product with numerical analysis and/or computer aided simulations before going into final production or manufacturing of the product. This is the same in the case of additive manufacturing too. But, additive manufacturing or 3D printing being a new technology, there are not yet enough industrially available simulation software packages that can predict all the steps and patterns related with this process in a comprehensive manner.

In the effort to create a comprehensive performance prediction tool, an intermediate analysis tool is needed to be developed, which can use the results available from currently existing simulation tools, match the results with experimental tests, analyze the effects of different correlated factors and finally provide a comprehensive and more accurate prediction of the process.

To predict the performance or behavior of a final product, it is possible to follow two different methods:

1. Using numerical modeling.
2. Using experimental testing on full scale or scaled down prototypes

13.2.1 Predict Using Numerical Modeling

This is the method that uses numerical simulations, FEA modeling of different types to predict different types of behavior of the mold and the final product. Currently, there are different simulation tools, but they are not comprehensive for 3D printed molds. So, we can do a number of various types of simulation, do some experimental testing, and then find out a trendline on how the simulation results matches or varies from experimental output and if there is a trend on how that varies with some related factors, and if there is some correlation between the result variation and those factors. All these results can finally lead to a more accurate prediction of mold behavior.

An example of prediction using numerical modeling can be shown from published research work of the authors [1]. This includes a simple transient thermal simulation on a plastic injection mold and predicting the total cooling time. The simulation predicts that in industrial setup, a plastic bottle cap made of polypropylene will take 28.25 s to cool down to 50 °C. While in the experimental testing within the actual industrial circumstances, the cooling time is measured to be 30s. Though this less than 2 s difference seem to be very insignificant, but it is very crucial for industry in mass production. It is necessary to establish a technique that reduces the 5% variation in numerical and simulation results for the industries to compete in todays business world.

13.2.2 Predict Using Prototypes

The aforementioned prediction model in the previous paragraph has an inherent difficulty within itself. To create an accurate trendline between experimental and simulation results, we need to create the full-size 3D printed molds and test them in actual industrial size injection molding machines, these are highly expensive, both the full size printed molds and also the molding machines. To minimize this difficulty, the easier prediction tool utilizes a smaller scale molding machines, small or scaled down molds and respective simulation tools. If anyone has a simulation model for analyzing larger molds, it's quite easy to analyze the smaller size molds too. But the difference comes in the cost savings in terms of the physical tools. This prediction technique involves identifying the experimental and numerical results of the scaled down mold, comparing them to the full-size molds numerical and experimental results, and trying to find out whether or not there are relations between full scale and scaled down models. It is expected to find a trend on how the smaller mold behaves and how the larger mold behaves and so on. The ultimate goal is to create such a tool that can predict the behavior of full size industrial mold from a smaller size simulation and/or in house experimental results.

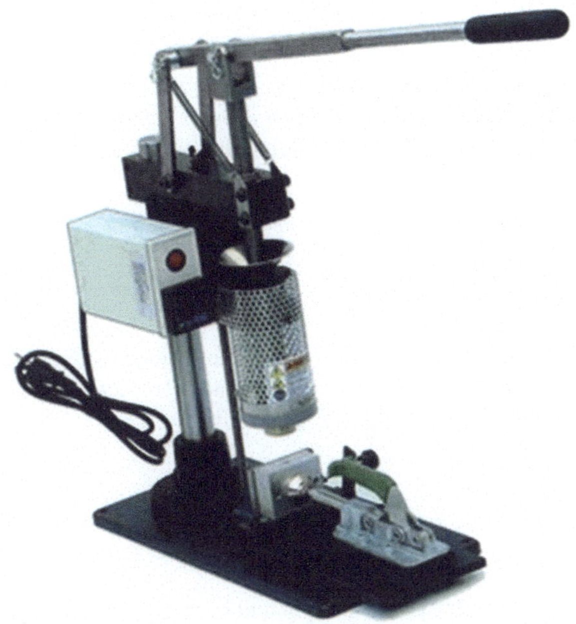

Fig. 13.1 Mini injection molding machine

Using a desktop mini injection molding machine, as shown in Fig. 13.1, it is possible to experimentally test a number of scaled down 3D printed molds. In the following sections, we will try to demonstrate a framework for design of experiments that uses these techniques and finally can provide accurate predictions of mold performances.

This paper is about creating the framework. Our future publications will include all the results that we can obtain using the prediction models. But we have some initial results, that can help to identify and illustrate the actual working model of the ongoing project.

13.2.3 Design of Experiments Framework

In the previous section, we have demonstrated just the basics of creating the comparison baseline, that can pave the way of a more comprehensive analysis framework. For this, we need a lot more results, both experimental and numerical. A number of Design of Experiments (DOE) are planned to establish a comprehensive analysis and prediction tool. The details of the design of experiments are mention below.

13.2.3.1 Design of Experiment 1

Purpose Identify the effect of mold material.

This DOE is designed to identify if tough resin plastic prototypes can serve as a substitute for stainless steel prototype. And if, we also need to find a trendline on the extent to which the correlation can be established. For this design of experiments, we are printing 4 sets of molds (one set of molding tool consists of two main parts; the cavity and the core).

1. Stainless steel mold for design 1(bottle cap).
2. Tough resin mold for design 1(bottle cap).
3. Stainless steel mold for design 2(fishing bait).
4. Tough resin mold for design 2(fishing bait).

These four molds are of identical size and shape externally, maximum longitudinal dimension being 3 inch on one side. They can fit into a simple manual desktop molding machine and product polypropylene plastic parts. These molds have straight cooling channels inside for better cooling and part quality. While running the actual molding experiments, the cooling rates can be varied (at least three different flow rate of coolant) to obtain large number of data sets to ensure maximum accuracy. Hence the total number of data points would be (4 × number of different coolant flow rate).

Benefit With sufficient number of data sets, this DOE can help to identify if 3D printed tough resin molds (low cost) can replace high cost 3D printed stainless steel ones. It will reduce the overall cost of experimental printing and predict the behavior and performance pattern of highly expensive full size industrial level stainless steel molds.

13.2.3.2 Design of Experiment 2

Purpose Identify the effect of mold size.

We already have a full size industrial level steel mold. It is needed to be tested against some other scaled down versions. Scale it down to two other sizes, such as 75%, 50%, and 25%. As a matter of fact, the first mold in DOE 1 (steel mold for design 1) is the 50% version here. The molds required in DOE 2 are:

1. 100% size stainless steel mold for design 1
2. 75% size stainless steel mold for design 1
3. 50% size stainless steel mold for design 1
4. 25% size stainless steel mold for design 1

After we have all four molds, we will test them both experimentally and numerically. From previous studies, it is known to the authors that the experimental and numerical results do not match well [2–7]. There is about 5–10% difference between them. It is required to find if there is a trend on how the variation works, and if there is a correlation between the sizing. For example, is it possible to find, say, the variation is 5% in full size molds, but 15% in 25% scaled down molds.

Benefit The benefits are twofold. First, by identifying how different the simulation results are from the experimental results, we can predict how the actual industrial results will be for other different molds in real life scenario. Second, if we can identify a trend between the variations at different sizes, we can just predict the performance of full size mold by from just experimental and simulations on very smaller size molds. This tremendously decreases the production cost and design and development stage cycle time.

13.2.3.3 Design of Experiment 3

Purpose Identify the effect of cooling channels.

This DOE is to see if a simulation tool can predict accurately how much does the cooling channel affects the cycle time and overall production cycle in injection molding. Currently existing injection molding tools are typically manufactured using traditional milling, drilling, grinding etc. operations, and most importantly cannot contain any sort of conformal cooling channels. Conformal channels have better cooling potentials for molding tools can be incorporated into mold by 3D printing process. To identify what type of cooling channel can serve the purpose best, we will design and analyze the following molds in the same overall size:

1. Stainless steel mold for design 1; no cooling channel.
2. Stainless steel mold for design 1; straight cooling channel.
3. Stainless steel mold for design 1; conformal cooling channel type1.
4. Stainless steel mold for design 1; conformal cooling channel type2.
5. Tough resin mold for design 1; no cooling channel.
6. Tough resin mold for design 1; straight cooling channel.
7. Tough resin mold for design 1; conformal cooling channel type1.
8. Tough resin mold for design 1; conformal cooling channel type2.

These molds can be tested both experimentally and numerically. Moreover, in numerical analysis we can run a number of tests with varying designs of conformal channels size, shape and cross section and hence identify the optimum design.

Benefits This DOE will provide in-depth information on the optimum cooling technique for injection molds. With sufficient resource and technique time, it is possible to analyze very different design and options of cooling and provide design specific optimal design rules for the mold designers.

13.2.3.4 Design of Experiment 4

Purpose Identify the effect of coolant fluid.

Experimentally we can use only a few coolants due to time and resource limitations. But while applying computer aided engineering, it is possible to test a number of coolants. The basic idea is to run experiments and simulations for limited number of molds (4) and find a trend on how the results relate and then based on other simulations with other coolants, predict the behavior and performance of those coolants to be used in practical industrial setup.

Following three molds will be used for experimental testing, with compressed air and water, hence 6tests in this DOE.

1. Stainless steel mold for design 1; straight cooling channel.
2. Stainless steel mold for design 1; conformal cooling channel type1.
3. Stainless steel mold for design 1; conformal cooling channel type2.

Benefits This DOE provides significant information on the properties, application, advantages and disadvantages, feasibility and behavioral predictability of different number of coolants in injection molding, which cannot be tested in practical scenario. Once we can establish that one fluid is optimum and feasible in certain condition, it becomes much easier from the business perspective to take any investment decision, such as changing cooling system in the industry.

13.2.3.5 Design of Experiment 5

Purpose Identify the effect of plastic material (the one for the final product).

This DOE is based on traditional simulation technique. To predict the behavior of a final product, it is common practice to create a simulation of the same and test it. For example, if a polypropylene plastic ball needs to be manufactured using industrial plastic injection molding machine, we should first simulate the same conditions in a computer aided simulation scenario and see how it works and take significant design decisions based on that. So, in this DOE a number of popular a highly consumer plastic materials are tested in simulation scenario and predict their behavior in industrial level. Moreover, with the help of inhouse desktop molding machine, we can use an intermediate prediction tool to estimate the products performance more accurately. Hence this DOE calls for both experimental and numerical techniques. The following molds are used for experimental study:

1. Stainless steel mold for design 1 (bottle cap).
2. Tough resin mold for design 1 (bottle cap).
3. Stainless steel mold for design 2 (fishing bait).
4. Tough resin mold for design 2 (fishing bait).

Benefits They is similar to the ones achieved in DOE. In place of coolants, it predicts the behavior of plastic product. And can demonstrate the performance of many different plastic types without doing any inhouse experiments and hence proves to be highly cost effective and less time consuming.

13.3 Conclusion

Although this paper does not contain any concrete technical results, it provides a working principle and framework for design of experiments which is expected to be very helpful for the plastic injection molding industry. The authors are currently working on the project and will discuss the results and findings of their work in future publications. With the limitation of available simulation tools and no source of data of actual industrial level usage of 3D printed injection molds being available yet, these DOEs will be able to provide significant information and suitable guidelines for the mold designers and molding industry.

References

1. Jahan, Suchana Akter.: Optimization of conformal cooling channels in 3D printed plastic injection molds. Diss. Purdue University (2016)
2. Wu, T., et al.: A framework for optimizing the design of injection molds with conformal cooling for additive manufacturing. Procedia Manuf. **1**, 404–415 (2015)
3. Jahan, S.A., El-Mounayri, H.: Optimal conformal cooling channels in 3D printed dies for plastic injection molding. Procedia Manuf. **5**, 888–900 (2016)
4. Jahan, S.A., et al.: Implementation of conformal cooling & topology optimization in 3D printed stainless steel porous structure injection molds. Procedia Manuf. **5**, 901–915 (2016)
5. Jahan, S.A., et al.: Thermo-mechanical design optimization of conformal cooling channels using design of experiments approach. Procedia Manuf. **10**, 898–911 (2017)
6. Wu, T., et al.: Design optimization of plastic injection tooling for additive manufacturing. Procedia Manuf. **10**, 923–934 (2017)
7. Jahan, S.A., et al.: Effect of porosity on thermal performance of plastic injection molds based on experimental and numerically derived material properties. In: Mechanics of Additive and Advanced Manufacturing, vol. 9, pp. 55–63. Springer, Cham (2018)

Chapter 14
Effect of Processing Parameters on Interlayer Fracture Toughness of Fused Filament Fabrication Thermoplastic Materials

Devin J. Young, Cara Otten, and Michael W. Czabaj

Abstract Additive manufacturing by fused filament fabrication (FFF) is a promising method for rapid manufacturing of complex components for a wide variety of applications. FFF is often limited to non-structural and non-load bearing applications due to insufficient strength and stiffness of the end-material. This is particularly true in the direction of layer deposition, due to poor adhesion between FFF layers. Processing parameters such as extrusion temperature and print speed have been shown to have significant effect on the mechanical performance of FFF components, but these studies have often neglected interlayer properties. This work develops and experimental approach for quantifying the relationship between processing parameters and interlayer fracture toughness of FFF specimens. The processing parameters considered include extrusion and bed temperatures, extrusion speed, raster spacing, and cooling-fan speed. FFF test blocks were fabricated to identify which parameters would best optimize interlayer fracture toughness. To measure interlayer fracture toughness, unidirectional ABS double cantilever beam specimens were fabricated according to the parametric test matrix with guidance from the test block results. In situ full-field thermography was used to record the specimen thermal history during fabrication. X-ray computed tomography was used to determine the internal void resulting from varying the raster spacing. Finally, optical and SEM fractography was used to perform post mortem categorization of specimen fracture surfaces. The fracture toughness data measured in this study is used to develop an approach for rapid optimization of interlayer properties of FFF components.

Keywords Additive manufacturing · Fused filament fabrication · Fracture toughness · Optimization · ABS

14.1 Introduction

Fused filament fabrication (FFF) is a common additive manufacturing (AM) process used to fabricate complex components for a variety of purposes. FFF parts are often unsuitable for load carrying applications due to reduced tensile and compressive properties when compared to other fabrication methods [1–4]. It has been shown that optimization of processing parameters has been successful in increasing mechanical performance of FFF parts [4–6].

Interlayer fracture properties have been largely ignored in the literature. A recent study has presented a method for the measure of interlayer fracture toughness for FFF specimens [7]. Among the conclusions of this study was the recognition that FFF parts have a reduced fracture toughness when compared to a neat resin specimen of the same material. It is unknown how various combinations of processing parameters effect interlayer fracture toughness.

This work documents a parametric study performed to determine the optimal combination of processing parameters to maximize interlayer fracture toughness. A test matrix was created considering five processing parameters. Single parameter permutation test blocks were created to lighten the testing burden by eliminating the least significant parameter options.

14.2 Experimental Methodology

Five processing parameters were considered in the test matrix: extrusion and bed temperatures, fabrication speed, raster spacing, and cooling fan speed. Three options are considered for each parameter. The test matrix is presented in Table 14.1.

D. J. Young · C. Otten · M. W. Czabaj (✉)
Department of Mechanical Engineering, University of Utah, Salt Lake City, UT, USA
e-mail: m.czabaj@utah.edu

S. Kramer et al. (eds.), *Mechanics of Additive and Advanced Manufacturing, Volume 8*, Conference Proceedings of the Society for Experimental Mechanics Series, https://doi.org/10.1007/978-3-319-95083-9_14

A test block specimen was designed to understand the effect of each parameter on interlayer adhesion. The test specimen is a unidirectional rectangular block 1 cm × 1 cm × 6 cm. Rasters are parallel to the longest dimension. A small triangular notch added to one end of the specimen to direct a traveling wedge between mid-plane layers and induce crack growth. A razor blade was swiped across the notch to produce a sharper crack front.

A baseline for processing parameters was established based on the parameters used for ABS specimens used in [7]. Test blocks were fabricated using these baseline parameters or with a modification to a single parameter. Specimens have been fabricated using the baseline parameters (marked in green in Table 14.1), a 65% infill, 225 °C extrusion temperature, and 245 °C extrusion temperature. Three blocks were manufactured for each parameter set.

Specimens were tested on a custom traveling wedge load frame designed in house at the Utah Composites Laboratory. The test blocks were held in a grip on one end while the wedge split the test block at the mid-plane. Applied load was measured with a load cell in line with the direction of wedge travel.

14.3 Results and Discussion

Wedge loading of the test blocks specimens yielded a consistent loading behavior: a linear elastic portion leading up to a maximum load value, followed by a reduction in load as a crack propagated through the midplane of the specimen. Specimens of each parameter set showed consistent loading and fracture behavior, reaching comparable peak loads (Fig. 14.1a). Each parameter set provided distinct loading and fracture behavior compared to other parameter sets (Fig. 14.1b). Variation of a single parameter (i.e. extrusion temperature) also yields distinct loading and fracture behavior (Fig. 14.1c).

Of note is that all specimens have similar slopes of the initial loading curves, suggesting a consistent elastic stiffness. The area under the loading curve is assumed to relate to the interlayer adhesion and thus interlayer fracture toughness. Better adhesion suggests more energy is required to cause layer separation. Thus, a higher maximum load can be associated with a higher fracture toughness value. Figure 14.1b lends credence to this theory as the baseline specimen was fabricated at a higher

Table 14.1 Test matrix for processing parameter study. It is expected that four specimens be produced for each parameter combination. Baseline parameters are marked in green.

Extrusion temp (°C)	Bed temp (°C)	Extrusion speed (mm/s)	Raster spacing (infill %)	Cooling fan speed (%)
225	75	30	65	0
235	95	45	80	50
245	105	60	100	100

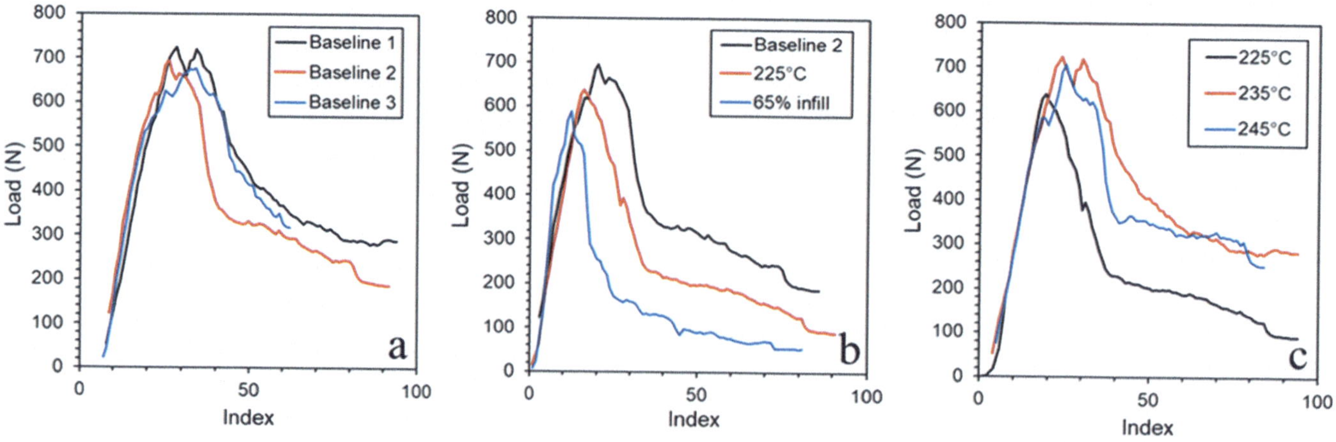

Fig. 14.1 Test block loading curves: (**a**) Overlaid loading curves for three baseline test blocks; (**b**) Comparison of baseline, 225 ° C, and 65% infill test blocks; (**c**) comparison of test blocks fabricated with different extrusion temperatures

temperature or greater infill percentage than the specimens it is compared to. Higher extrusion temperature provides more time for polymer chain entanglement, increasing adhesion. Greater infill percentage indicates more material bonded between layers and requires more energy to separate them. The use of these test blocks provides a promising method of determining which parameters will be most important for maximizing interlayer fracture toughness. Consideration of the test block results will help reduce the testing burden by eliminating the processing parameters that have the least effect on fracture toughness.

14.4 Conclusions

Qualitative assessments of interlayer fracture toughness have been made using unidirectional test blocks fabricated from ABS filament. Fracture in the test blocks has been induced using a custom load frame with a traveling wedge to split the test blocks at the mid-plane. Test blocks varied one from another by the permutation of a single processing parameter. Test blocks showed a consistent loading and fracture behavior. Mechanical response varied distinctly between blocks with different processing parameters. Test blocks will be fabricated for the remaining parameter options. The results of the test blocks will provide insight into which processing parameters have the greatest effect on fracture toughness and provide guidance for a parametric study on fracture toughness of FFF specimens.

References

1. Smith, W.C., Dean, R.W.: Structural characteristics of fused deposition modeling polycarbonate material. Polym. Test. **32**(8), 1306–1312 (2013)
2. Zaldivar, R.J., et al.: Influence of processing and orientation print effects on the mechanical and thermal behavior of 3D-printed ULTEM® 9085 material. Addit. Manuf. **13**, 71–80 (2017)
3. Koch, C., Van Hulle, L., Rudolph, N.: Investigation of mechanical anisotropy of the fused filament fabrication process via customized tool path generation. Addit. Manuf. **16**, 138–145 (2017)
4. Ahn, S.-H., et al.: Anisotropic material properties of fused deposition modeling ABS. Rapid Prototyp. J. **8**(4), 248–257 (2002)
5. Tymrak, B.M., Kreiger, M., Pearce, J.M.: Mechanical properties of components fabricated with open-source 3-D printers under realistic environmental conditions. Mater. Des. **58**, 242–246 (2014)
6. Sood, A.K., Ohdar, R.K., Mahapatra, S.S.: Parametric appraisal of mechanical property of fused deposition modelling processed parts. Mater. Des. **31**(1), 287–295 (2010)
7. Young, D., Wetmore, N., Czabaj, M.: Interlayer fracture toughness of additively manufactured unreinforced and carbon-fiber-reinforced Acrylonitrile Butadiene Styrene. Additive Manufacturing, accepted for publication (2018)
8. ASTM International, ASTM D5528–13, Standard test method for mode I interlaminar fracture toughness of unidirectional fiber-reinforced polymer matrix composites, West Conshohocken, PA (2013)

Chapter 15
Forced-Response Verification of the Inherent Damping in Additive Manufactured Specimens

Onome Scott-Emuakpor, Tommy George, Brian Runyon, Bryan Langley, Luke Sheridan, Casey Holycross, Ryan O'Hara, and Philip Johnson

Abstract The laser powder bed fusion AM process has been used to manufacture beams with unique internal geometries that are capable of increasing inherent damping in a part. The concept of the internal design is to have densely packed, unfused powder pockets that dissipate energy via particle interaction. Four Inconel (IN) 718 beams have been tested and all demonstrated the capability to suppress vibration 10X more effectively than a fully fused beam. The mechanism presumed to dissipate energy and thus suppress vibration is the sliding of unfused particles. This mechanism has been associated with a crack opening under Mode II fracture. Based on this assumption, a proportional expression has been developed as a criterion for optimizing unfused powder locations for vibration suppression effectiveness and was validated with 3.175 mm thick beams. This study investigates five uniquely designed IN-718 beams created via the optimizing criterion to assess accuracy of the expression. The intent of this study is to investigate the predictability of the unfused pocket optimization criterion. The results of this study will lead to a more robust design criterion for more complex 3D structures with improved damping capability.

Keywords Damping · Resonance · Fatigue · Additive manufacturing · Laser powder bed fusion · Inconel 718 turbine engine component design

15.1 Introduction

Gas turbine engine performance can be improved by weight reduction of rotating componentry. Weight reduction in modern and next generation gas turbine engines are realized by replacing inserted airfoil and disk combinations (more often seen in legacy fleets) with integrally bladed rotors (IBRs) which weigh significantly less. Though IBRs are great for engine performance, the absence of airfoil-to-disk energy dissipation via friction reduces system damping and increases the susceptibility of IBR airfoils to vibration. This makes high cycle fatigue (HCF) a more prevalent issue in IBRs and could lead to catastrophic engine failure. Over the past two decades, the devastation caused by HCF has fostered changes in critical component design guidelines to avoid the phenomenon. When unavoidable, however, post-design as well as operational measures are considered for the component to sustain HCF loads [1]. Vibration suppression techniques have been sought out as viable approaches that can be used either during or post IBR design to combat HCF issues. Investigated for a number of years, vibration suppression studies have yielded unique techniques such as viscoelastic treatment, hard coatings (ceramic- and metallic-based), as well as impact and friction dampers [2–13]. Each method has been proven effective; however, all have short comings that make them less than ideal (though not totally impractical) for turbine engine applications. Some of these short comings include weight addition and lower damage tolerances.

A recent vibration suppression option without the short comings mentioned in modern damping techniques was recently introduced [14]. The approach uses the laser powder bed fusion (LPBF) additive manufacturing (AM) process to fabricate beams with unique internal geometries, where the internal features are capable of increasing system damping and reducing vibration. The concept of the internal design is to have densely packed, unfused powder pockets that dissipate energy via

O. Scott-Emuakpor (✉) · T. George · B. Runyon · B. Langley · L. Sheridan · C. Holycross
Air Force Research Laboratory, Dayton, OH, USA
e-mail: onome.scott-emuakpor.1@us.af.mil

R. O'Hara
Air Force Institute of Technology, Dayton, OH, USA

P. Johnson
Universal Technology Corporation, Dayton, OH, USA

S. Kramer et al. (eds.), *Mechanics of Additive and Advanced Manufacturing, Volume 8*, Conference Proceedings of the Society for Experimental Mechanics Series, https://doi.org/10.1007/978-3-319-95083-9_15

particle interaction. Four Inconel (IN) 718 beams with unique internal designs (only 1–4% of the beam volume was unfused powder) have been tested and all demonstrated the capability to increase damping (i.e. suppress vibration) by 10X the magnitude of a fully fused beam. Further investigation of these beams highlighted particle sliding as the mechanism of damping, and a theory associating this motion to a crack opening under mode II fracture allowed for a criterion for optimizing internal unfused geometries [14]. This study investigates the accuracy of the criterion for optimizing unfused geometry locations by conducting forced-response tests on IN-718 beams with strategically placed internal pockets. A key finding in the results highlight a need for a more robust design criterion independent of manufacturing parameters and specific dimensional changes. The following sections describe the experimental approach, the beam designs, and the results of the theoretical and empirical data comparisons.

15.2 Experimental Procedure

Forced-response tests are conducted on cantilevered Inconel 718 beams with the five dimensions illustrated in Fig. 15.1; note, the beam thickness is 6.35 mm. The internal pocket dimensions and locations illustrated in Fig. 15.1 are determined via the optimization criterion which is proportional to Mode II fracture energy release rate: i.e. the product of shear squared and crack motion. This criterion was developed based on the assumption that the powder inside the unfused pocket is densely packed. In other words, inside the unfused powder pocket, there is no room for stochastic particle motion [14]. The purpose of the designs in Fig. 15.1a–d are to have beams with maximum and minimum damping influence at 2nd and 3rd bend, and the two pocket beam (Fig. 15.1e) has maximum damping influence at 2nd bend and marginal influence (compared to max) at 3rd bend. The results of the optimization criterion value with respect to specific beam designs in Fig. 15.1 are plotted in Fig. 15.2. In the figure, a higher value means the beam will have more damping [14].

The excitation for the vibration test is supplied with a 27 kN Unholtz-Dickie electrodynamic shaker [11]. The shaker is controlled with a Vibration Research VR5000 controller, and an Endevco 2271A charge-type accelerometer is used to monitor shaker head acceleration and control voltage supple. Another instrumentation used during testing is a Polytec OFV 505 single-point laser vibrometer for velocity frequency response measurement of the test article. The velocity response from the laser vibrometer is the data used to assess damping behavior. The data is captured via a swept sine test at a rate greater than 2 bandwidths per minute (with respect to the half power bandwidth method) for a frequency range spanning at least three bandwidths. Damping capability is then calculated as quality factor (Q) via the half power bandwidth method [15]. The image in Fig. 15.3 shows the setup of the described forced-response experiments.

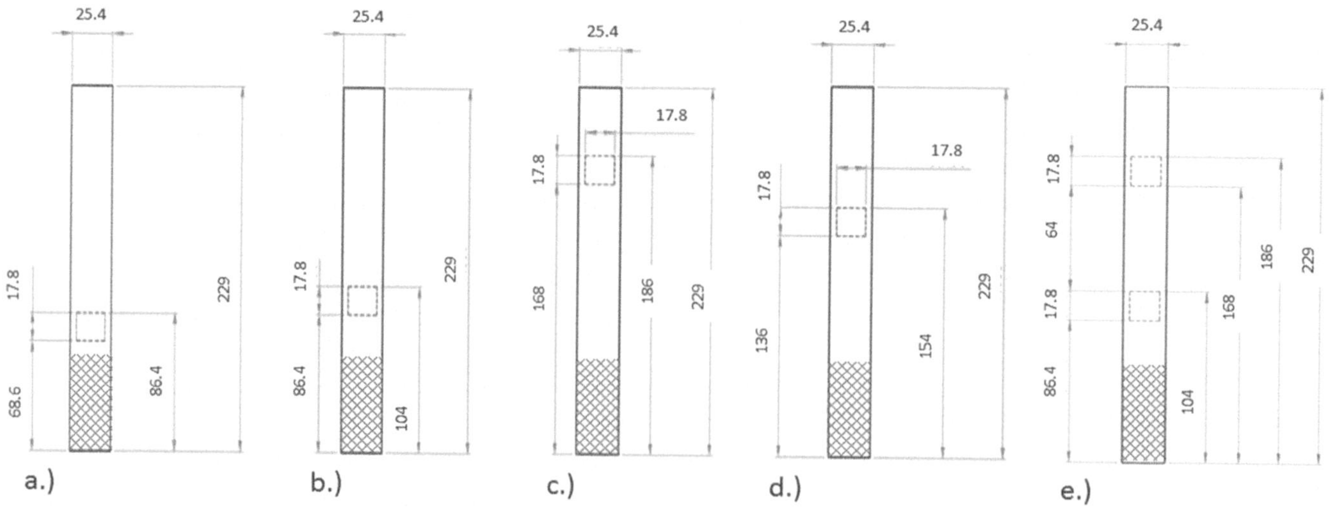

Fig. 15.1 Five unique beam with unfused powder pockets, cross-hatched area is 51 mm inside the clamp and the beam free length is 178 mm: (**a**) 3rd MAX, (**b**) 2nd MAX, (**c**) 3rd MIN, (**d**) 2ND MIN, and (**e**) Two pocket

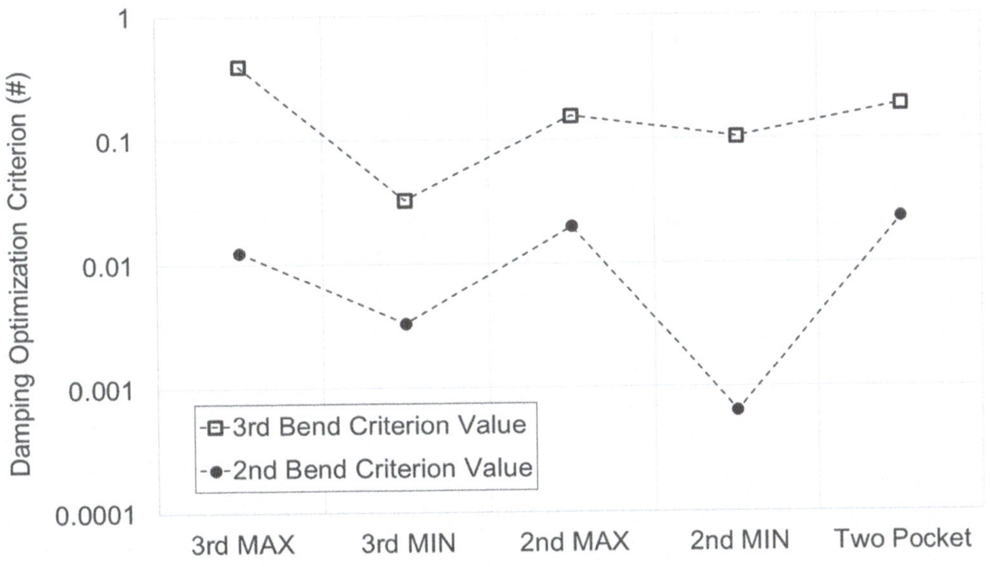

Fig. 15.2 Damping optimization criterion results: higher number means more damping capability

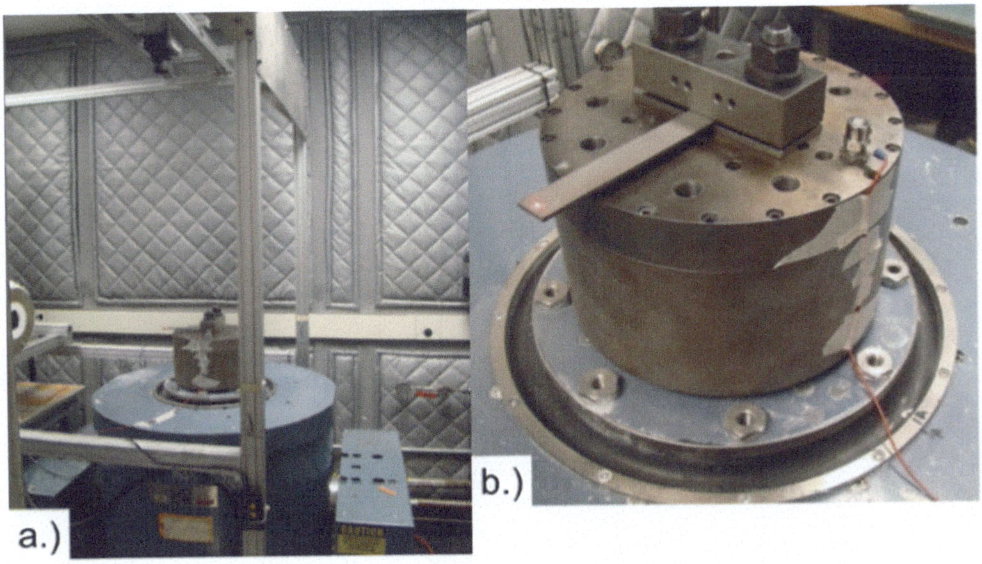

Fig. 15.3 Shaker test setup with cantilever clamped beam: (**a**) 27 kN shaker, (**b**) test-setup with beam and tower

15.3 Results

Five beams mentioned in the experimental procedure were excited with a base excitation of 0.1 G (gravitational constant 9.81 m/s). The results for Q values calculated by the half power bandwidth method are listed in Table 15.1. Comparisons of these Q values to the damping optimization criterion values from Fig. 15.2 is illustrated in Fig. 15.4. Results of the comparison show that the optimization theory does not support the Q values attained from the optimally designed beams. An immediate justification could point to the thickness difference between the beams tested in this study (6.35 mm) versus previous efforts (3.175 mm) [14]. However, a deeper investigation is required to verify that the key physics of the optimization theory holds true.

Table 15.1 Damping results for five beams with unfused powder pockets

Beam #	Bending Mode	Frequency (Hz)	Q
2ndMIN	3rd bend	2543.4	585
	2nd bend	921.9	105
	1st bend	147.7	295
2ndMAX	3rd bend	2436	70
	2nd bend	864.5	211
	1st bend	791.6	317
3rdMIN	3rd bend	2467	77
	2nd bend	872.7	485
	1st bend	141.5	283
3rd MAX	3rd bend	2432	116
	2nd bend	857	429
	1st bend	139.7	466
Two pocket	3rd bend	2442	64
	2nd bend	844.1	156
	1st bend	139.6	465

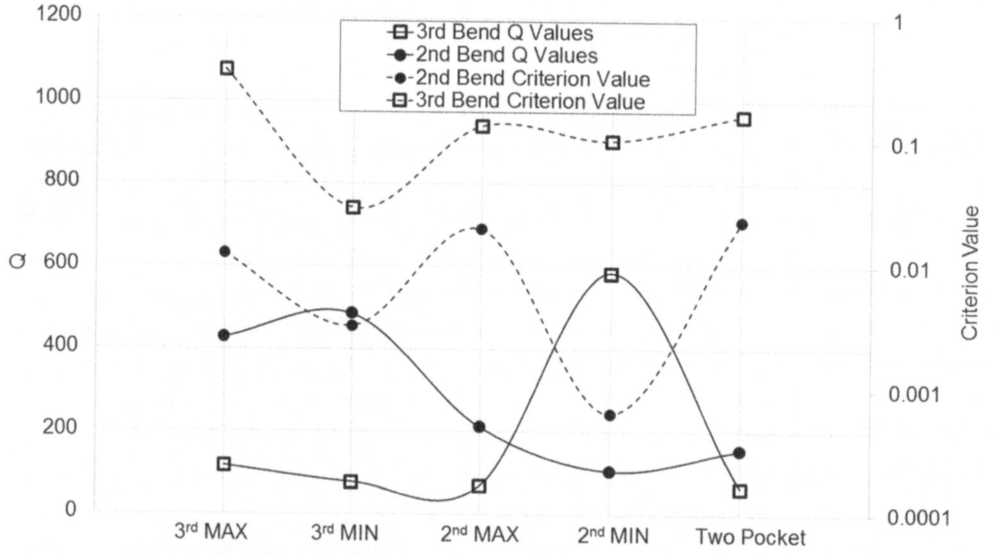

Fig. 15.4 Comparison of beam damping capability (Q) and the optimization criterion values

Mentioned in previous sections regarding the optimization theory identifying pocket placement for superior damping is that the pocket be densely packed with unfused powder. This statement is critical to minimizing the stochastic nature of particle motion and associating the remaining, known motion to Mode II crack opening. Based on the criticality of densely packed powder pockets, each beam is scanned using computed tomography (CT) and the internal characteristics are analyzed, specifically at the location of pockets. With a 14 μm voxel, images like the one in Fig. 15.5 are generated and analyzed. The image in Fig. 15.5 clearly shows that the unfused powder pocket is not densely packed: i.e. empty space, denoted by red outline. This revelation means that the optimization theory does not apply to the tested beams because motions other than particles sliding (similar to Mode II crack opening motion) can occur. The result of particle motion alternative to that of the prior developed optimization theory is unpredictable inherent energy dissipation, and thus unpredictable inherent damping capability.

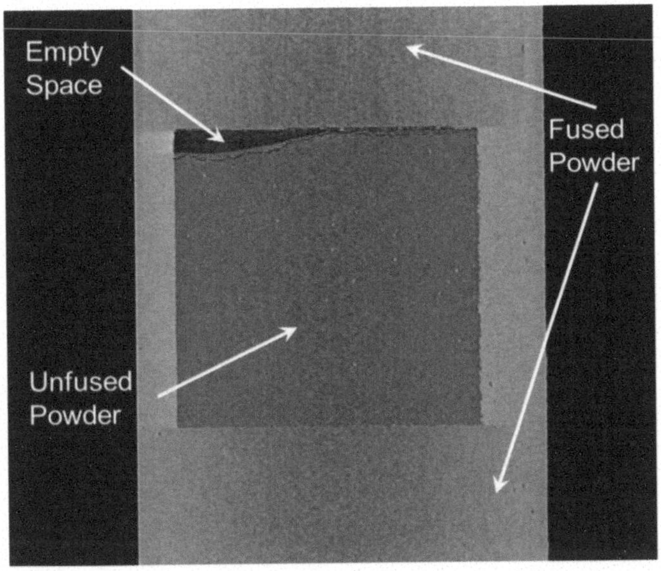

Fig. 15.5 CT scan of an unfused internal pocket of a beam: darker image means no density, thus no powder

15.4 Future Work

Results of this study highlighted a limitation in the applicability of the damping optimization criterion for specimens with unfused powder pockets. Based on the results, several investigations are anticipated for future efforts. First, a generalized or secondary criterion dealing with partially packed powder pockets with be developed. Second, strain amplitude effects on damping capability and repeatability for beams that are not densely packed will be investigated; specifically, beams in this study will be observed closely. Third, a detailed study of the powder application and packing density using the LPBF process will be conducted. Influences such as strain amplitude and specimen thickness on damping capability are currently under investigation.

Acknowledgements The authors would like to thank the Turbine Engine Fatigue Facility (TEFF) of the United States Air Force Research Laboratory (AFRL) and Universal Technology Corporation (UTC) for funding, support, and collaboration. Specifically, the authors would like to acknowledge UTC contractor Philip Johnson, Angela Still, Thaddeus Crowe, Ross Cefalu, and Christopher Howard for contributing to laboratory testing in the TEFF.

References

1. Engine Structural Integrity Program (ENSIP), MIL HDBK-1783B (USAF), 15 Feb 2002
2. Nashif, A.D., Jones, D., Henderson, J.P.: Vibration Damping. John Wiley, New York (1985)
3. Torvik, P.J., Patsias, S., Tomlinson, G.R.: Characterizing the damping behaviour of hard coatings: a comparison from two methodologies. Proceedings of the 7th National Turbine Engine High Cycle Fatigue Conference, May 2002, West Palm Beach, FL
4. Olson, S.: An analytical particle damping model. J. Sound Vib. **264**(5), 1155–1166 (2003)
5. Lopez, I., Busturia, J., Nijmeijer, H.: Energy dissipation of a friction damper. J. Sound Vib. **278**(3), 539–561 (2004)
6. Jones, D., Parin, M.: Technique for measuring damping properties of thin viscoelastic layers. J. Sound Vib. **24**(2), 201–210 (1972)
7. Panossian, H.: Structural damping enhancement via non-obstructive particle damping technique. ASME J. Vib. Acoust. **114**(1), 101–105 (1991)
8. Torvik, P.: Damping properties of hard coatings for engine applications. Adv. Sci. Tech. **66**, 126–135 (2010)
9. Reed, S.: Development of experimental, analytical, and computational techniques appropriate for nonlinear damping coatings. Ph.D. Dissertation, Dept. of Aeronautics and Astronautics, Air Force Institute of Technology, Wright-Patterson AFB (2007)
10. Torvik, P.J.: On estimating system damping from frequency response bandwidths. J. Sound Vib. **330**(25), 6088–6097 (2011)
11. Scott-Emuakpor, O., Langley, B., Holycross, C., George, T., Runyon, B., Justice, J.: Comparison Between Forced-Response and Hysteretic Energy Damping Assessment Methods, pp. 4–8. AIAA Science and Technology Forum and Exposition, San Diego (January 2016)
12. Torvik, P., Langley, B.: Properties of Hard Coatings with High Damping, pp. 27–29. AIAA Joint Propulsion Conference and Exposition, Orlando (2015)

13. Torvik, P., Wilson, R., Hansel, J.: Influence of a Viscoelastic Surface Infiltrate on the Damping Properties of Plasma Spray Alumina Coatings, Part I: Room Temperature," proceedings of the Materials Science and Technology Conference and Exhibition, Detroit, MI, 2007, pp. 139–150
14. Scott-Emuakpor, O., George, T., Runyon, B., Holycross, C., Langley, B., Sheridan, L., O'Hara, R., Johnson, P., Beck, J.: Investigating Damping Performance of Laser Powder Bed Fused Components with Unique Internal Structures. ASME/Turbo Expo, Oslo, Norway, 11–15 June 2018, paper No. GT2018–75977
15. American Society for Testing and Materials, E756–05: Standard Test Method for Measuring Vibration-Damping Properties of Materials, ASTM Book of Standards, 2010; Vol. 04.06, ASTM International, West Conshohocken, PA

Chapter 16
Computational and Experimental Characterization of 3D Printed Components by Fused Deposition Modeling

Koohyar Pooladvand and Cosme Furlong

Abstract This paper presents the development of methodologies to understand the effects of process parameters in 3D printed components' performance and geometrical characteristics, specifically distortions and residual stresses. Full-field-of-view noninvasive optical metrology methodologies and computational simulations outline the framework of this approach. We are developing computational models to predict the critical attributes of 3D printed parts by Fused Deposition Modeling (FDM). We are also designing particular testing artifacts with specific shapes and geometries to conduct Non-Destructive Testing (NDT) using full-field-of-view optical sensors, i.e., Digital Holographic Interferometry, Digital Image Correlation, and Digital Fringe Projection. These sensors can be utilized during and after fabrication for extraction of mechanical properties, identification of defects, and characterization of geometrical accuracies/distortions as a function of process parameters. The knowledge gained from NDT results is used for tuning our computational models. Representative results demonstrate the feasibility of the proposed computational-experimental approach for potential implementation into FDM processes in order to understand the interconnection between process parameters and part performance, which eventually will lead to improvements in the integrity, repeatability, and consistency of printed components and to reduced costs and optimized energy consumption.

Keywords 3D printing · Computational modeling · Finite Element Method (FEM) · Fused Deposition Modeling (FDM) · Non-destructive testing · Non-invasive optical sensors

16.1 Introduction

One of the promising polymer-based 3D printing technologies that has been widely adopted is Fused Deposition Modeling (FDM). In spite of its rapid development in recent years, polymer-based Additive Manufacturing (AM) is still suffering from critical limitations including rough surface finish, uncertain dimensional accuracy, and precision, as well as low mechanical strength and reliability of the printed components [1, 2]. Turning FDM into a leading producer of functional components requires addressing these limitations.

In polymer-based AM, e.g., FDM, neither experimental nor numerical studies have solely been able to address the influences of process and manufacturing parameters on part performance. On the one hand, simulations have been used to understand the effects of process parameters on some characteristics of printed components, i.e., layer bonding, mechanical properties, distortions, microstructures, and residual stresses [1–5]; however, effective experimental verifications are still needed. On the other hand, well-developed nondestructive testing (NDT) procedures capable of measuring structural properties and health are available. Methods such as Digital Holographic Interferometry (DHI), Digital Image Correlation (DIC), and Digital Fringe Projection (DFP) enable characterization of mechanical properties, shape and geometrical distortions, as well as porosity and defect identification [6–9].

We are proposing a method to tackle the challenges in FDM using both numerical and experimental approaches concurrently. Our method consists of simulations that can capture the complex physics of 3D printing and can be tuned based on nondestructive measurements of thermal and structural characteristics. These measurements consist of thermograms, distortions, natural frequencies, and moduli of elasticity. Within our method, we are designing, simulating, printing, and testing artifacts to investigate how different printing and manufacturing parameters, including layer thickness, envelope

K. Pooladvand (✉) · C. Furlong
Center for Holographic Studies and Laser micromechaTronics – CHSLT, Mechanical Engineering Department, Worcester Polytechnic Institute, Worcester, MA, USA
e-mail: kpooladvand@wpi.edu

S. Kramer et al. (eds.), *Mechanics of Additive and Advanced Manufacturing, Volume 8*, Conference Proceedings of the Society for Experimental Mechanics Series, https://doi.org/10.1007/978-3-319-95083-9_16

temperature, extruder temperature, heated bed temperature, feed flow rate, extruder speed, rastering, and geometries affect the performance and quality of the artifacts. This will allow us to identify the mechanical and geometrical properties as a function of process parameters.

16.2 Methods

In this paper, we are particularly assessing applicability and capability of combined computational-experimental approaches to address challenges in 3D printed components. Figure 16.1 outlines our proposed methodology to eventually lead to process optimization and performance improvement. We are studying the thermally driven phenomena in 3D printing by FDM to understand the development of distortions and residual stresses in testing artifacts having specific shapes and dimensions. To do this, the transient thermal distributions and flows are initially estimated and subsequently applied as loads to transient non-linear structural simulations. We can predict the distortions and residual stresses by modifying the assumptions and settings made in the simulations after performing experimental verifications. Such assumptions include boundary conditions, extruder temperature, heat transfer coefficients, and material properties, as well as element types, mesh sizes, time step, and material types.

We utilize noninvasive sensors, e.g., thermography, and optical metrology methodologies, during and after printing to determine the temperature distributions, heat fluxes, distortions, and structural properties of 3D printed components. We design testing artifacts to be printed under different process parameters and configurations. These artifacts are aimed to respond predictably to excitations along their principal axes. We perform harmonic and impulse load excitations to carry out modal analysis and transient dynamic measurements, respectively. We conduct modal analysis by Digital Holographic Interferometry (DHI) for recording the natural frequencies and corresponding mode shapes. Moreover, the transient dynamic measurements are made by applying different impulse excitations to the artifacts and measuring their time-dependent vibrations for further analyses. These analyses enable the determination of structural properties such as elastic moduli, damping coefficients, stiffness, and shear moduli, as well as deviations from the normal response that might point to microstructural defects [7, 9–12]. These properties are related to different printing and manufacturing conditions. In addition, distortions on printed components are measured employing Digital Fringe Projection (DFP) that helps to understand the interconnections between parameters and the status of the printed components regarding residual stresses, distortions, and after-fabrication evolutions.

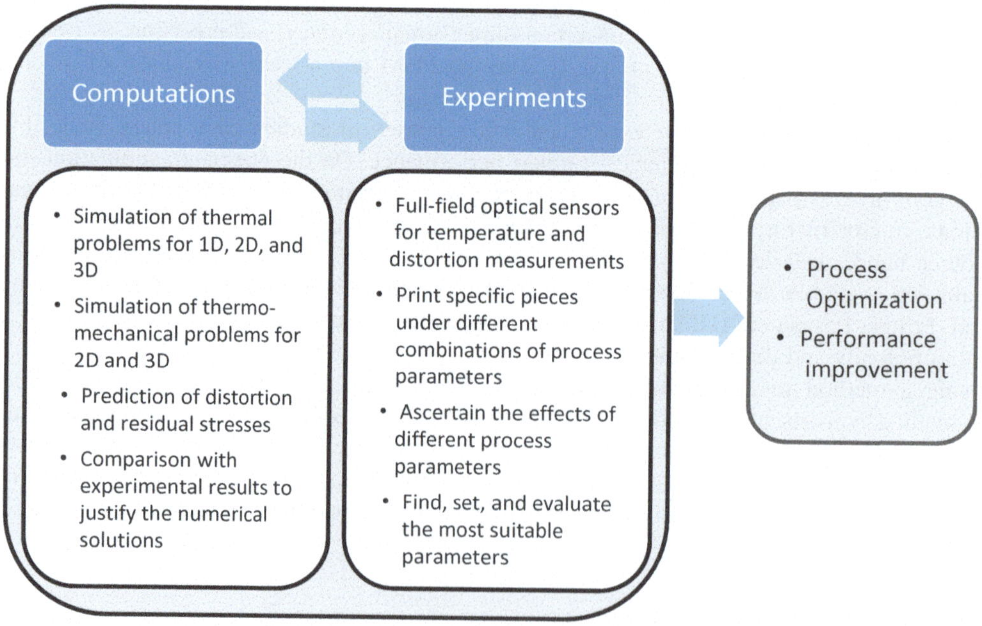

Fig. 16.1 Framework of the combined computational-experimental methodology to study FDM processes. Methodology will enable research toward understanding process parameters and part performance

16.2.1 Thermo-Mechanical Modeling and Estimation of Distortions and Residual Stresses

Simulations are done by numerically solving partial differential equations of two different physics: Eq. 16.1 for a nonlinear transient thermal [1, 13], and Eq. 16.2 for a nonlinear structural [14],

$$\frac{\partial \rho u}{\partial t} + V.\nabla \rho h = \nabla.(\kappa \nabla T) + \dot{q}, \tag{16.1}$$

$$\Delta\varepsilon_t^{tot} = \Delta\varepsilon_t^{e} + \Delta\varepsilon_t^{th} + \Delta\varepsilon_t^{p}, \tag{16.2}$$

where t is time, u is internal energy, ρ is mass density, V is velocity vector, h is enthalpy, κ is the thermal conductivity, T is temperature, and $\dot{q}$ is volumetric heat generation. In Eq. 16.2, ε is strain tensor, superscript e indicates elastic displacement, p plastic, th thermal, and tot total strain., and subscript t is time. Δ indicates the incremental change in each component of the strain fields.

Figure 16.2 illustrates our approach for simulation of the deposition in Cartesian coordinates. Deposition starts by progressively laying sets of elements in the directions shown in Fig. 16.2a. For example, as in Fig. 16.2b, deposition starts when the extruder moves from right to left on the first layer and reversely on the next layer above. This process is repeated until the final layer is deposited after which the model continues in order to simulate the cooling processes and the removal of the part from the heated bed located underneath the first layer.

The half-length of the beam is modeled by assuming symmetry with respect to the y-z plane, which is located at the origin. Models consist of two different physics, thermal and structural, with the same geometry but different corresponding material properties and element types. The general geometry with the applied boundary conditions is shown in Fig. 16.2c. Representative results, Fig. 16.3a, b, show the contours of temperatures and deformations at different times during simulations. In these simulations, the part is not rigidly attached to a heated bed at the bottom and acts similar to a cantilever beam fully constrained at one end.

We also included contact elements to simulate bonding between the heated bed and the printed component. Considering this bonding enables one to simulate the peeling of the part from the heated bed due to the accumulation of residual stresses and distortions. Figure 16.4 shows representative results illustrating the prediction of distortions and peeling as well as stress distributions in three directions. We intend to use simulations to verify and compare distortions between predictions and those in actual printed components.

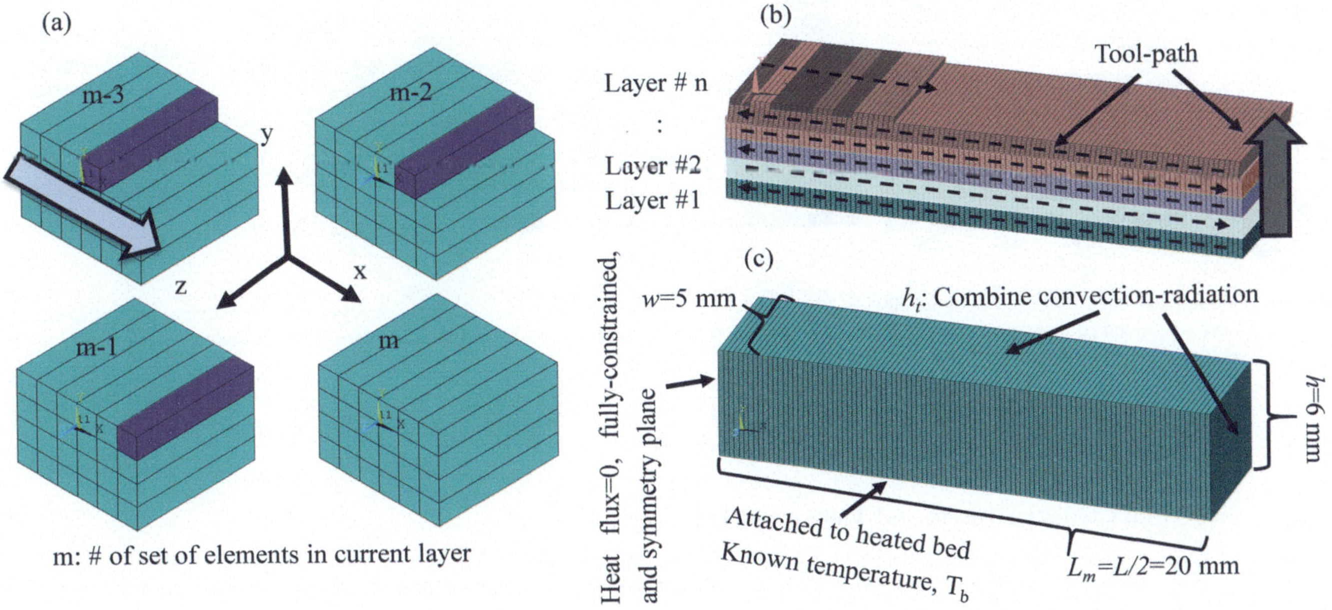

Fig. 16.2 3D finite element and schematic representation of the layer-by-layer deposition strategy: (**a**) sequential deposition of sets of elements within a current layer, with each active set (in purple) laid in time increment dt; (**b**) sequences of bottom-up deposited layers simulated as built up in a 3D printer; and (**c**) representative boundary conditions for thermal and structural analyses

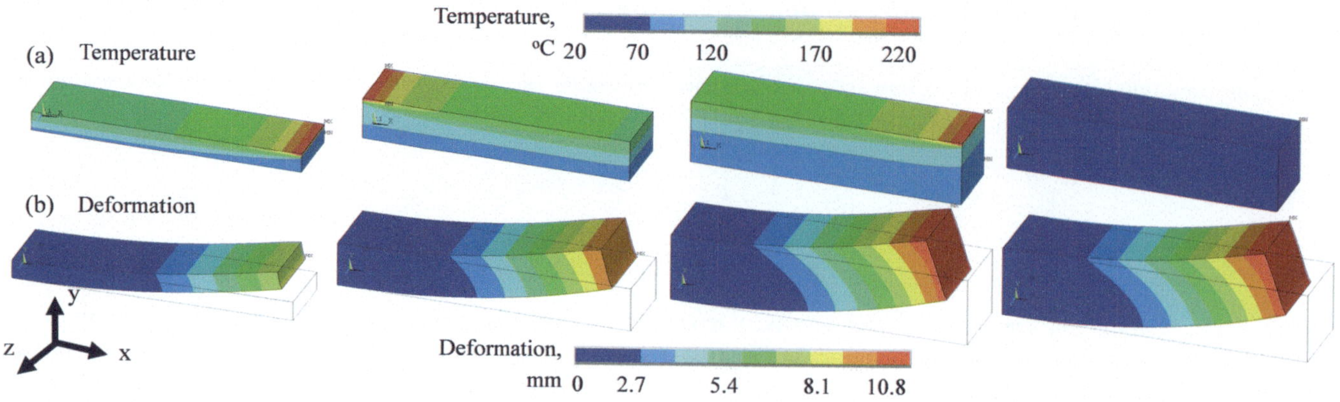

Fig. 16.3 Representative computational results corresponding to four instances during manufacturing of a beam including 33% completion, 66% completion, the instance of laying the last set of elements, and the moment part reaches room temperature, respectively from left to right: (**a**) temperature distributions; and (**b**) resultant deformations

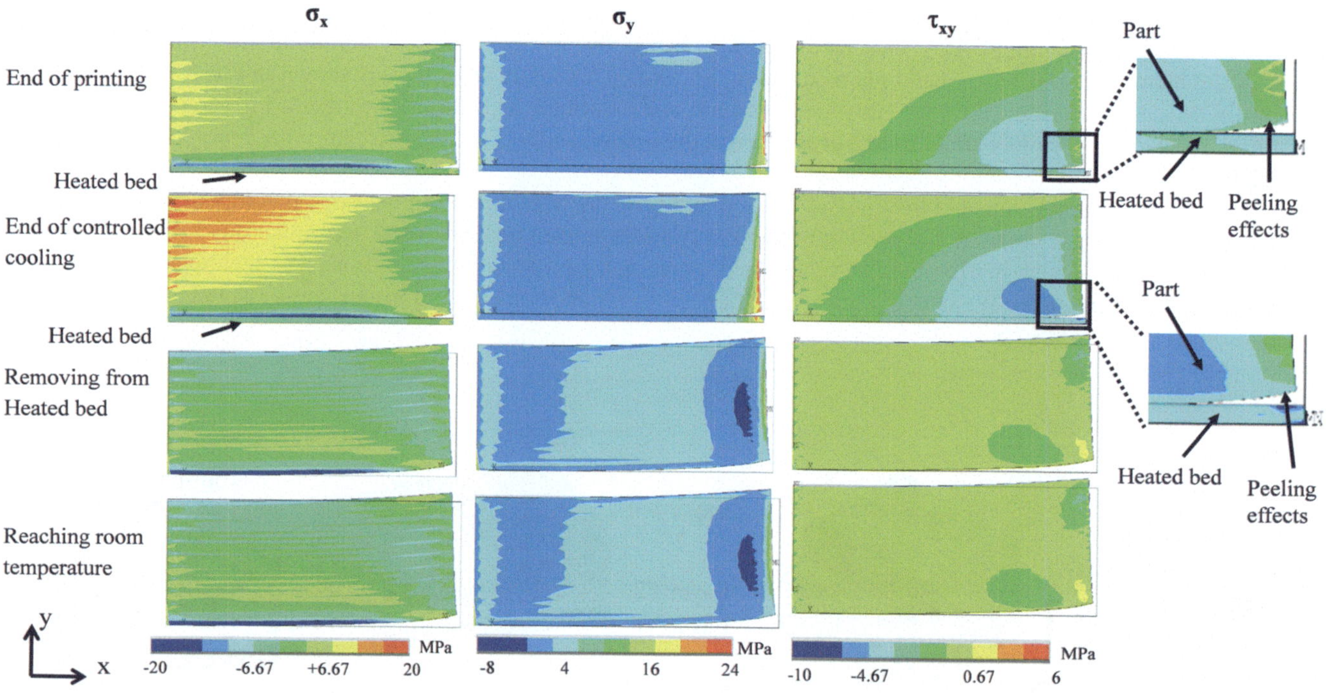

Fig. 16.4 Computationally predicted contour plots of three principal residual stresses, σ_x, σ_y *and* τ_{xy} at four critical instances, including end of printing, end of controlled cooling attached to the heated bed, removal from the heated-bed, and after reaching room temperature. Simulations can be used to predict peeling of the part from the heated bed

16.2.2 Experimental Measurements and NDT

16.2.2.1 Thermal Characterization and Tuning by IR Imaging

An Infrared (IR) camera shown schematically in Fig. 16.5a is used to measure the surface temperature and capture the thermograms *in-situ*. We use these measurements for tuning our developed numerical models as described in [13]. For example, Fig. 16.5b represents an achievable agreement between estimated and measured temperatures for different instances during fabrication along a vertical line on the surface of a cylinder with diameter and height of 5.0 ± 0.3 mm and 50 ± 0.09 mm, respectively.

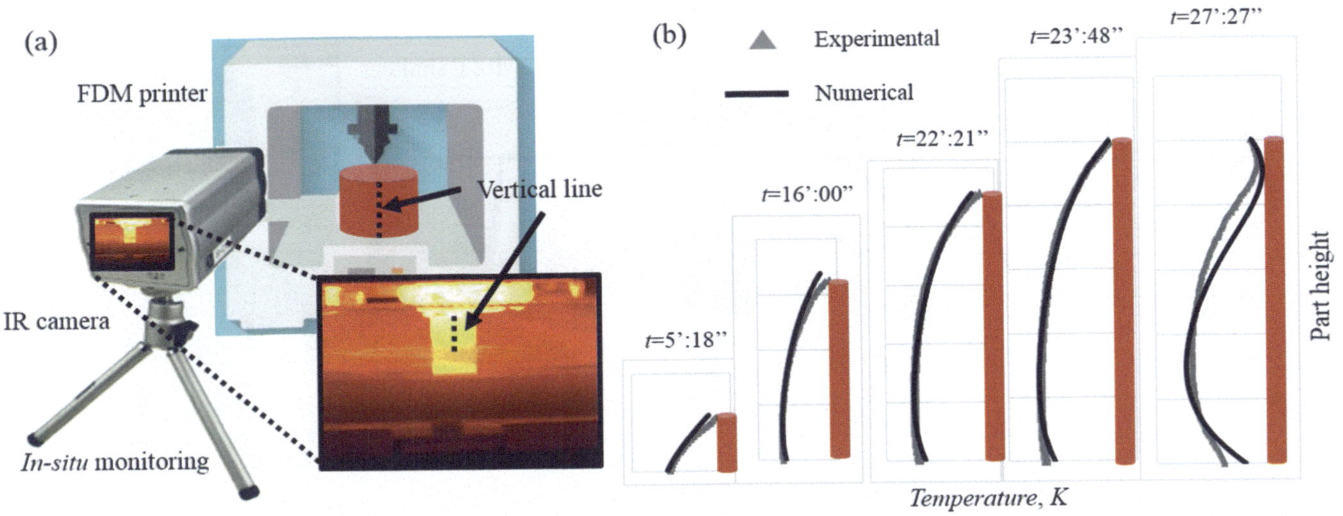

Fig. 16.5 (**a**) Schematic of the IR camera monitoring the fabrication processes *in-situ*; and (**b**) representative results of the agreement between numerically estimated and experimentally measured temperatures along a vertical line on a cylinder with diameter of 5.0 ± 0.3 mm and height of 50 ± 0.09 for different instances

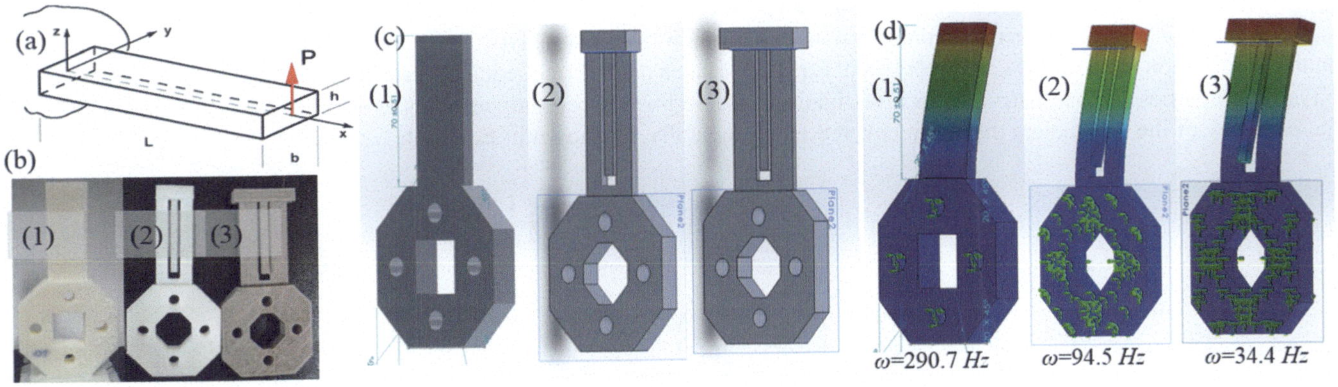

Fig. 16.6 Representative designed testing artifacts for NDT: (**a**) schematic of a basic cantilever design; (**b**) three different printed testing artifacts based on (**a**); (**c**) solid models of the designs; and (**d**) modal analyses showing the first bending modes of (**c**)

16.2.2.2 Design of Testing Artifacts

We use the concept of a cantilever beam to design testing artifacts with geometries similar to those shown in Fig. 16.6a–c. We parametrize our designs such that it is possible to alter their geometrical, structural, and physical properties from one to another. For each of them, we utilize computational models using finite element method to obtain their corresponding natural frequencies and mode shapes. Representative estimations of the first bending modes for three selected designs are illustrated in Fig. 16.6d. We relate the geometry and shape to structural characteristics and subsequently study the effects of specific process parameters through experimental investigations by modal and impulse excitation analyses.

16.2.2.3 Structural Analysis by NDT Testing

Testing artifacts similar to the ones shown in Fig. 16.6b are used to characterize mechanical and structural properties of components printed with different process parameters. High-speed 3D Digital Image Correlation (HS-DIC) and Digital Holographic Interferometry (DHI), similar to ones shown in Fig. 16.7a, b, respectively, are used for these purposes.

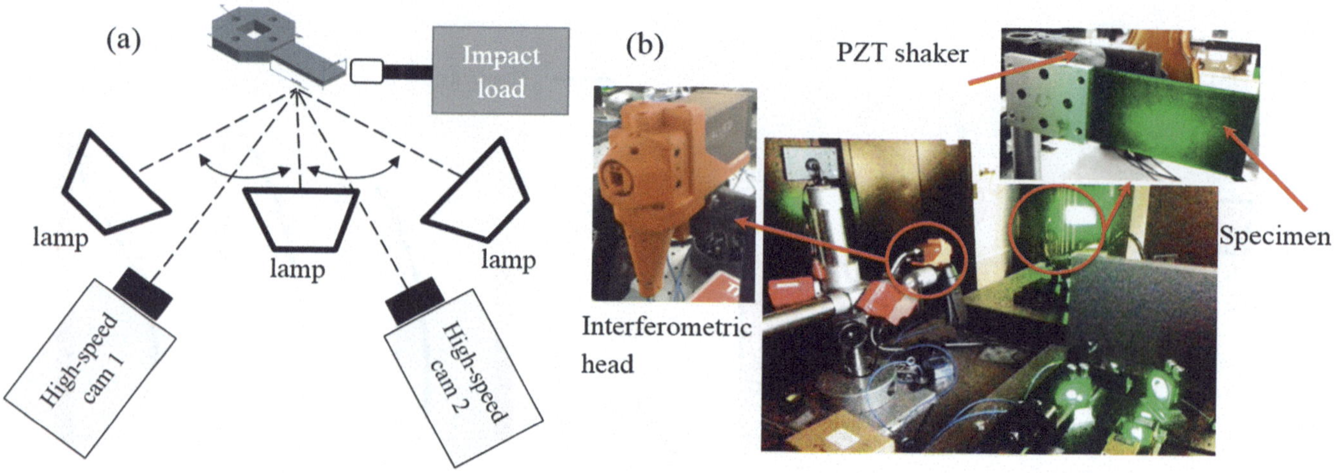

Fig. 16.7 Experimental setups for NDT of testing artifacts: (**a**) 3D DIC with ultra high-speed stereo cameras, lighting, imaging optics and excitation apparatus; and (**b**) lensless DHI equipped with high-frequency piezoelectric shakers

Testing artifacts are studied under steady state and transient loading conditions. Steady-state responses are investigated with DHI while the load is applied with a piezoelectric shaker by sweeping the excitation frequencies in order to identify resonance conditions. Transient investigations are performed with HS-DIC, which is setup to capture the high-speed displacements of the testing artifacts due to impact loads in axial and flexural directions. These two loading conditions yield such properties as moduli of elasticity, Poisson's ratio, and damping that correlate with printing parameters. Due to the unique design of the testing artifacts, their responses are predictable by analytical and numerical solutions. For instance, an analytical solution for a simple cantilever beam, Eq. 16.3, relates natural frequencies to the moduli of elasticity as well as to its physical properties and geometry. In our investigations, we perform uncertainty analysis to obtain a level of understanding on the accuracy of the parameters that are recovered experimentally, as shown in Eq. 16.4.

$$E = E(\rho, \omega_n, L, h) = \rho\left(\frac{12 \times \omega_n \times L^2}{\beta \times h^2}\right)^2, \tag{16.3}$$

$$\delta E = \left\{\left(\delta\rho \times \frac{\partial E}{\partial \rho}\right)^2 + \left(\delta\omega_n \times \frac{\partial E}{\partial \omega_n}\right)^2 + \left(\delta L \times \frac{\partial E}{\partial L}\right)^2 + \left(\delta h \times \frac{\partial E}{\partial h}\right)^2\right\}^{\frac{1}{2}}, \tag{16.4}$$

where, ω_n is natural frequency, ρ is density, L is the length, and h is height. Constant β is 1.015 for the first bending mode.

Furthermore, noninvasive measurements of shapes and distortions are critical to our method, and we perform instantaneous as well as time-lapsed measurements by Digital Fringe Projection (DFP). These measurements provide information of deformations and distortions of components during and after fabrication.

16.3 Representative Results

We performed measurements of distortion of 3D printed components similar to the one shown in Fig. 16.8a, which shows that after printing, the part continues to deform due to stress relaxations. Measurements illustrated in Fig. 16.8b began immediately after printing of the object and recorded distortions under laboratory conditioned for 10 h using Digital Fringe Projection (DFP). The deformation shows an average exponential behavior, which is to be investigated as a function of process parameters.

To illustrate the variability of the mechanical properties as a function of process parameters, we printed samples similar to those shown in Fig. 16.6b. Two process parameters were varied, including printing speed as well as printing orientation. Representative results, shown in Table 16.1, correspond to a cantilever beam similar to design (1) of Fig. 16.6b with length, width, and thickness of 70 ± 0.25 mm, 30 ± 0.25 mm, and 6 ± 0.09 mm, respectively. The corresponding natural frequencies

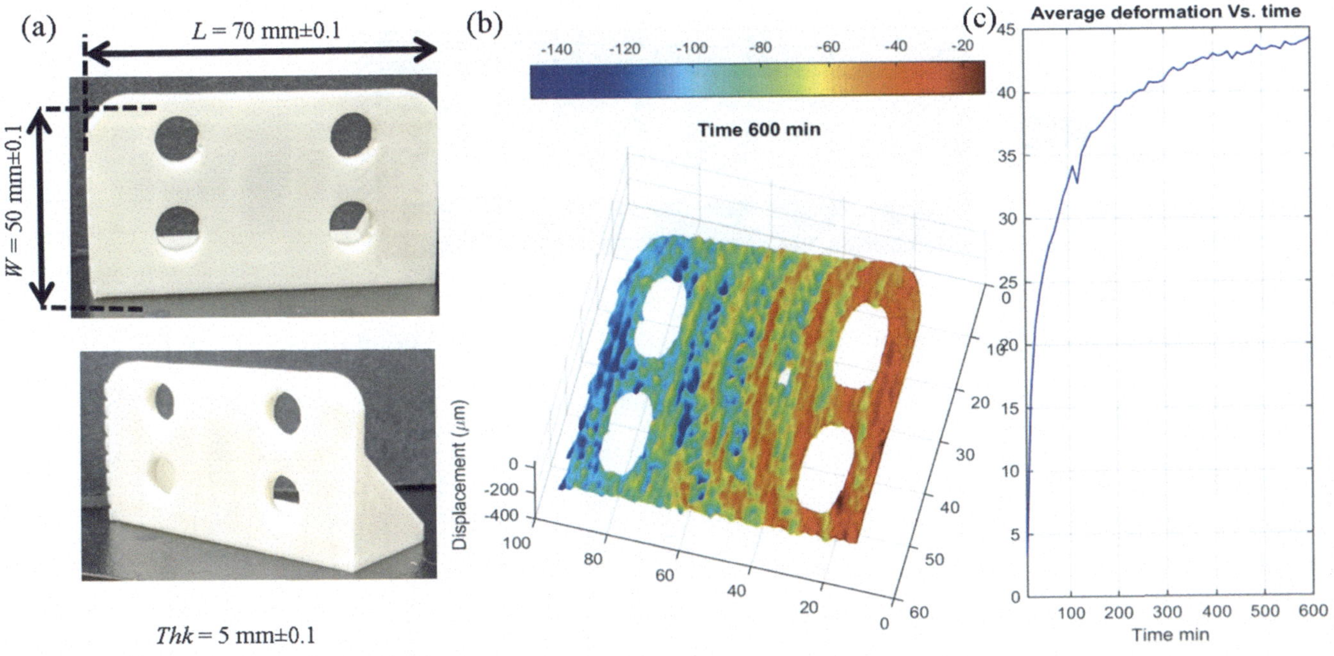

Fig. 16.8 Monitoring of the out-of-plane deformation on a component for 600 min using DFP: (**a**) 3D printed component; and (**b**) measured out-of-plane deformation on the front surface of the sample; and (**c**) averaged out-of-plane displacement

Table 16.1 Recovered moduli of elasticity of testing artifacts when fabricated at varying printing speeds and orientations. Experimentally obtained first modes of vibration were utilized. Pristine ABS filament with a reference module of elasticity of 2030 MPa was used to compute deviations

Printing speed	Printing direction	1st bending ω_n (Hz)	Module of elasticity (MPa)	Deviation from Ref., %
Fast	Horizontal	299 ± 5	2148.3	+5.83%
	Vertical	270 ± 5	1751.8	−13.71%
Normal	Horizontal	294 ± 5	2077.1	+2.32%
	Vertical	258 ± 5	1599.53	−21.21%
Slow	Horizontal	291 ± 5	2034.9	+0.24%
	Vertical	223 ± 5	1195.0	−41.13%

were measured by DHI, and the moduli of elasticity were recovered using Eqs. 16.3 and 16.4. Further, deviations were calculated relative to the properties of pristine ABS filaments.

As illustrated in Table 16.1, results show that the printing speed is significantly important. This corresponds to the effect of energy deposition as reported by others [1, 2, 15]. Outcomes confirm the positive effect of printing speed (feed rate) on improving the bonding between layers. Horizontal printing demonstrates elasticity close to the pristine ABS filament because of the alignment of the filament with the axial direction of the beam; but still, the effects of better bonding due to the higher energy deposition associated with higher feed rate are noticeable.

We also conducted impact studies on the same 3D printed part using HS-DIC. Representative results presented in Fig. 16.9 show the corresponding transient response for impact in x and y directions, respectively. These data can be evaluated using Fourier analysis to obtain the natural frequencies and can be used to assess the health and the porosity of the part nondestructively.

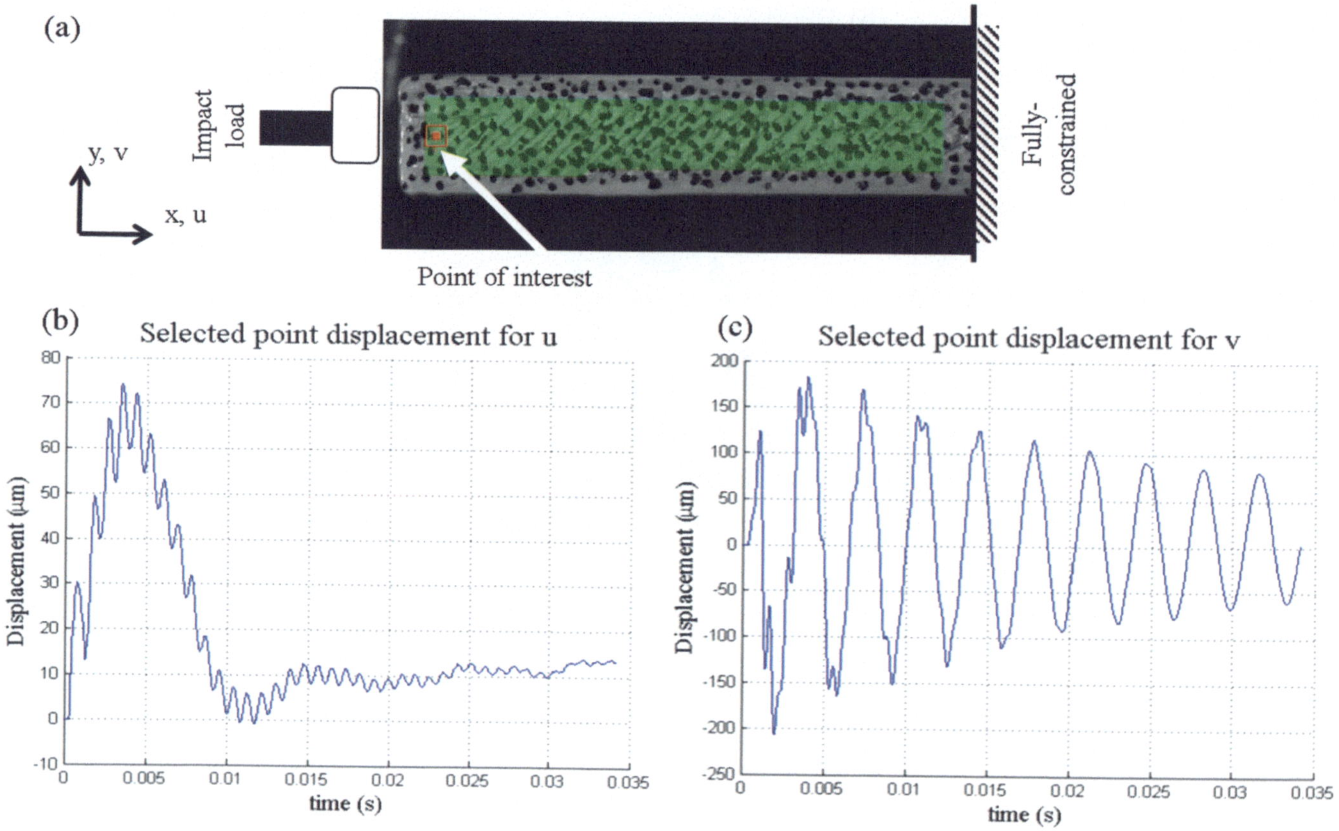

Fig. 16.9 Representative results of axial impact on a 3D printed component by HS-DIC: (**a**) side view of the artifacts prepared for DIC measurements also showing the location of the impactor; (**b**) displacements in *x*-direction, *u*, for 35 msec; and (**c**) displacements in *y*-direction, *v*, for 35 msec. These measurements will be utilized to study structural health of the part as well as part porosity

16.4 Conclusions and Future Work

In this paper, we presented our efforts on the development of computational-experimental methodologies to investigate process parameters in 3D printing by FDM and their relationship to components' performance. Computational simulations estimate the evolution of distortions and residual stresses in components printed by FDM during and after fabrication. A few boundary conditions and settings are assumed to be tunable and adjustable in the simulations according to feedbacks from experimental investigations. Optical methodologies are employed for modal, impact, and distortions analyses of the 3D printed components with unique designs to recover structural properties and provide feedbacks for simulations.

The proposed methodologies benefitting from nondestructive testing by optical methodologies and computational simulations can contribute to extending our understanding beyond the current level to improve the integrity, repeatability, and consistency of 3D printing processes such as FDM.

Future work includes the development of more refined testing artifacts to elucidate the interrelationships between process parameters and part performance. Further, applicability of specific nondestructive testing methodologies to interrogate parameters such as porosity, defect detections, as well as residual stresses and distortions will be explored, and findings will be presented in the future.

Acknowledgments This work is being partially supported by NSF award CMMI-1428921. We would also like to gratefully acknowledge the support of the Mechanical Engineering Department of Worcester Polytechnic Institute (WPI) and contributions by members of the CHSLT.

References

1. Zhang, Y., Chou, Y.: Three-dimensional finite element analysis simulations of the fused deposition modelling process. Proc. Inst. Mech. Eng. B J. Eng. Manuf. **220**(10), 1663–1671 (2006)
2. Zhang, Y., Chou, K.: A parametric study of part distortions in fused deposition modelling using three-dimensional finite element analysis. Proc. Inst. Mech. Eng. B J. Eng. Manuf. **222**(8), 959–968 (2008)
3. Favaloro, A., Brenken, B., Barocio, E., DeNardo, N.M., Pipes, R.B.: Microstructural modeling of fiber filled polymers in fused filament fabrication. Proc. SAMPE Conference, Long Beach, CA (2016)
4. Brenken, B., Favaloro, A., Barocio, E., DeNardo, N.M., Pipes, R.B.: Development of a model to predict temperature history and crystallization behavior of 3D printed parts made from fiber-reinforced thermoplastic polymers. Proc. SAMPE Conference, Long Beach, CA (2017)
5. Favaloro, A.J., Brenken, B., Barocio, E., Pipes, R.B.: Simulation of polymeric composites additive manufacturing using Abaqus. Proc. Science in the Age of Experience Conference, pp. 15–18
6. Pooladvand, K., Furlong, C.: Digital holography and digital image correlation in additive manufacturing. ISEM 2015, 5th International Symposium on Experimental Mechanics, Guanajuato, Mexico (2015)
7. Digilov, R.M., Abramovich, H.: Flexural vibration test of a beam elastically restrained at one end: a new approach for Young's modulus determination. Adv. Mater. Sci. Eng. **2013**, 1–6 (2013)
8. Buchaillot, L., Farnault, E., Hoummady, M., Fujita, H.: Silicon nitride thin films Young's modulus determination by an optical non-destructive method. Japanese J. App. Phy. **36**(6B), L794 (1997)
9. Roebben, G., Bollen, B., Brebels, A., Van Humbeeck, J., Van der Biest, O.: Impulse excitation apparatus to measure resonant frequencies, elastic moduli, and internal friction at room and high temperature. Rev. Scient. Inst. **68**(12), 4511–4515 (1997)
10. Burdzik, R., Stanik, Z., Warczek, J.: Method of assessing the impact of material properties on the propagation of vibrations excited with a single force impulse. Arch. Metal. Mater. **57**(2), 409–416 (2012)
11. Zeng, D.-J., Zheng, Q.-S.: Resonant frequency-based method for measuring the Young's moduli of nanowires. Phys. Rev. B. **76**(7), 075417 (2007)
12. Sandia, N.L.: Non-Destructive Additive Manufacturing Characterization Coupon. Sandia National Laboratories. https://ip.sandia.gov (2018)
13. Pooladvand, K., Furlong, C.: Thermo-mechanical investigation of fused deposition modeling by computational and experimental methods. In: Proc. SEM, Mechanics of Composite and Multi-functional Materials, vol. 7, pp. 45–54. Springer. https://link.springer.com/chapter/10.1007/978-3-319-41766-0_6 (2017)
14. Kamara, A., Marimuthu, S., Li, L.: A numerical investigation into residual stress characteristics in laser deposited multiple layer waspaloy parts. *J. Manuf. Sci. Technol.* **133**(3), 031013 (2011)
15. Thomas, J., Rodríguez, J.: Modeling the fracture strength between fused deposition extruded roads. Proc. Proceedings of the 11th Solid Freeform Fabrication Symposium, pp. 16–23

Chapter 17
Linking Thermal History to Mechanical Behavior in Directed Energy Deposited Materials

Jian Cao

Abstract Additive manufacturing is a promising process that has the capability for process optimization and materials development of novel, multi-material and functional components of complex geometries due to the rapid and localized directional solidification of molten metallic alloys. Directed energy deposition, an additive manufacturing process that uses a high powered laser to melt blown metallic powder, introduces large gradients and sensitivity in thermal histories within a built component that lead to unique phase transformations, microstructures, residual stress and anisotropic mechanical behavior. Control of the overall mechanical behavior of DED-built components relies on control of thermal history at localized areas. Research at Northwestern University, in collaboration with Argonne National Laboratory, uses in-situ monitoring techniques such as infrared (IR) cameras, an IR two-wave pyrometer to monitor the melt pool, and a high-powered synchrotron to capture the phase change during build. Relationships between temperature, solidification rate and thermal gradient are made with the resulting microstructural characteristic and mechanical behavior at localized areas of each build. Linking thermal history to mechanical behavior of additively-built parts will lead to increased thermal control for optimal properties and open the door to alloy development.

Keywords Additive manufacturing · Metal · Directed energy deposition · Characterization

Additive manufacturing is a promising process that has the capability for process optimization and materials development of novel, multi-material and functional components of complex geometries due to the rapid and localized directional solidification of molten metallic alloys. Directed energy deposition, or DED, is an additive manufacturing process that uses a high-powered laser to melt blown metallic powder, and introduces large gradients and sensitivity in thermal histories within a built component that lead to unique phase transformations, microstructures, residual stress and anisotropic mechanical behavior [1, 2]. Because of simultaneous re-melting of underlying layers and surrounding areas during the process, the DED-processed component undergoes directional solidification with grain coarsening in the build direction with the exception of the top layer [3]. Another microstructural artifact that occurs due to directional solidification is porosity, with lack of fusion, irregularly shaped porosity at interfaces and spherically shaped, vaporized porosity within the melt pool [4, 5]. Control of the overall mechanical behavior of DED-built components relies on control of thermal history at localized areas.

In-situ monitoring techniques such as infrared (IR) cameras and high-speed synchrotron imaging are used to capture the temperature during build. Experimental techniques to capture temperature are coupled with thermal models, to better predict for the thermal history at localized areas. This paper will discuss two DED-processed materials and their relationships between processing conditions, thermal history, structure and the resulting mechanical behavior.

Three Ti-6Al-4V cubic components were built with different laser power inputs into the melt pool [6]. A thermal model was used to compare the solidification rate (the cooling rate within the solidus and liquidus temperatures of the titanium alloy) at localized areas. Process parameters and material properties were inputs into the thermal model. The solidification rates were compared to characteristics of experimentally-determined, localized porosity maps with a range of lack of fusion to vaporization porosity. Statistically representative relationships between solidification rate and porosity size, shape, and distribution led to predicted and meshed porosity maps and localized mechanical behavior. The localized mechanical behavior captured the anisotropy and heterogeneity of the DED-processed titanium cubes and how build, hatch and scan directions influence mechanical behavior at various orientations. Linking computationally-determined temperature with porosity

J. Cao (✉)
Department of Mechanical Engineering, Northwestern University, Evanston, IL, USA
e-mail: jcao@northwestern.edu

S. Kramer et al. (eds.), *Mechanics of Additive and Advanced Manufacturing, Volume 8*, Conference Proceedings of the Society for Experimental Mechanics Series, https://doi.org/10.1007/978-3-319-95083-9_17

structures and the resulting tensile behavior allowed for a framework that linked solidification at any point within a component and its resulting properties [6].

DED-processing of the nickel-based superalloy, INCONEL 718 (IN718), included surface thermal monitoring using an IR camera. Several single clads were built by varying the laser power and powder mass flow. Thermal histories of localized points of the single clad were compared to the various scales of the resulting structure, including clad height, dilution into the underlying substrate, dendritic spacing and Laves phases within the cross-section of the clad. Structural investigations were linked to the micro-hardness at various points of each clad, at various cross-sectional points. These results were used to calibrate and validate micro-hardness prediction models that use thermos-fluidic dynamic melt pool models to predict for structure and mechanical behavior.

IN718 single clads expanded to building of thin walls, or a build with 120 layers in the Z direction but no re-melting in the lateral directions. IR camera temperature readings of the surface were coupled with thermal models to capture temperature histories within the build as well. Dwell time, or the amount of time that allows for cooling of the component between each layer, was the key factor in the thin wall components. IN718 walls with no dwell time between each layer experienced an order of magnitude slower solidification cooling rates than components with a 1-min dwell time because the 1 min dwell component allows for complete cooling after each built layer. The large difference in thermal history led to differences in crystallographic structure and residual strain determined with micro-Laue X-ray diffraction, with greater residual strains in the 1-min dwell component. Differences in porosity structure between no dwell and 1-min dwell thin walls were captured by using X-ray computed tomography of machined tensile specimens. After initial scans, the tensile specimens underwent uniaxial load with several pauses for relaxation for tomography scans to capture the evolution of additively-manufactured induced porosity. This work captures the links between the process parameter of dwell time with porosity and crystallographic structure and tensile behavior.

The knowledge of how thermal history links with mechanical behavior of additively-built parts will pave the way for using thermal control to achieve desired properties and engineered materials. The next steps will be to incorporate models that predict microstructures and mechanical behavior into process design as a component is being built. High-speed synchrotron imaging during the DED process can capture the formation of defects, powder scattering and entrainment and solidification epitaxial lines in real time. The challenge of capturing physical phenomena during additive manufacturing is having the capability of high temporal resolution metrology for solidification rates of up to 10^5 K/s. However, the opportunity is to use this in-situ information to inform predictive models and control the structure during the process for defects or unique phase transformations.

Acknowledgements This works acknowledges the support by the National Institute of Standards and Technology (award no. 70NANB13H194), Center for Hierarchical Materials Design (CHiMAD, award no. 70NANB14H012), and Digital Manufacturing and Design Innovation Institute (DMDII, award no. 15-07). Northwestern University professors Kornel Ehmann, Wing Kam Liu, Gregory J. Wagner and students Sarah Wolff, Hao Wu, David Pritchet, Stephen Lin, Orion Kafka, Jennifer Bennett, Puikei Cheng, Cheng Yu and Haiguang Liu contributed to this work. This work made use of facilities at Northwestern University, Quad City Manufacturing Laboratory, DMG MORI and the Advanced Photon Source at Argonne National Laboratory.

References

1. Shamsaei, N., Yadollahi, A., Bian, L., Thompson, S.M.: An overview of direct laser deposition for additive manufacturing; part II: mechanical behavior, process parameter optimization and control. Addit. Manuf. **8**, 12–35 (2015)
2. Heigel, J.C., Michaleris, P., Reutzel, E.W.: Thermo-mechanical model development and validation of directed energy deposition additive manufacturing of Ti–6Al–4V. Addit. Manuf. **5**, 9–19 (2015)
3. Wang, Z., Palmer, T.A., Beese, A.M.: Effect of processing parameters on microstructure and tensile properties of austenitic stainless steel 304L made by directed energy deposition additive manufacturing. Acta Mater. **110**, 226–235 (2016)
4. Kobryn, P.A., Moore, E.H., Semiatin, S.L.: The effect of laser power and traverse speed on microstructure, porosity, and build height in laser-deposited Ti-6Al-4V. Scr. Mater. **43**(4), 299–305 (2000)
5. Wolff, S.J., Lin, S., Faierson, E.J., Liu, W.K., Wagner, G.J., Cao, J.: A framework to link localized cooling and properties of directed energy deposition (DED)-processed Ti-6Al-4V. Acta Mater. **132**, 106–117 (2017)
6. Wolff, S., Lee, T., Faierson, E., Ehmann, K., Cao, J.: Anisotropic properties of directed energy deposition (DED)-processed Ti–6Al–4V. J. Manuf. Process. **24**, 397–405 (2016)

The manufacturer's authorised representative in the EU is Springer Nature Customer Service Centre GmbH, Europaplatz 3, 69115 Heidelberg, Germany. If you have any concerns regarding our products, please contact ProductSafety@springernature.com

Printed and bound by CPI Group (UK) Ltd, Croydon, CR0 4YY
06/10/2026
02228820-0001